LEGAL ISSUES IN GLOBAL CONTEXTS: Perspectives on Technical Communication in an International Age

Edited by

Kirk St.Amant
East Carolina University

and

Martine Courant Rife
Lansing Community College

Baywood's Technical Communications Series
Series Editor: Charles H. Sides

Baywood Publishing Company, Inc.
AMITYVILLE, NEW YORK

Printed in the United States of America on acid-free recycled paper.

Baywood Publishing Company, Inc.

26 Austin Avenue
P.O. Box 337
Amityville, NY 11701
(800) 638-7819

E-mail: baywood@baywood.com
Web site: baywood.com

Library of Congress Catalog Number: 2013036928
ISBN: 978-0-89503-835-7 (cloth : alk. paper)
ISBN: 978-0-89503-836-4 (paper)
ISBN: 978-0-89503-837-1 (e-pub)
ISBN: 978-0-89503-838-8 (e-pdf)
http://dx.doi.org/10.2190/LII

Library of Congress Cataloging-in-Publication Data

Legal issues in global contexts : perspectives on technical communication in an international age / edited by Kirk St.Amant, East Carolina University and Martine Courant Rife, Lansing Community College.
pages cm. -- (Baywood's technical communications series)
Includes bibliographical references and index.
ISBN 978-0-89503-835-7 (cloth : alk. paper) -- ISBN 978-0-89503-836-4 (pbk. : alk. paper) -- ISBN 978-0-89503-837-1 (e-pub) -- ISBN 978-0-89503-838-8 (e-pdf) 1. Computers--Law and legislation--United States. 2. Intellectual property--United States. 3. Information technology--Law and legislation--United States. 4. Technological innovations--Law and legislation--United States. 5. Law and globalization. 6. Technology and law. I. St.Amant, Kirk, 1970- editor of compilation. II. Rife, Martine Courant, editor of compilation.
KF390. 5. C6L445 2013
343.09'99--dc23

2013036928

Table of Contents

FOREWORD

Considering Legal Issues in Global Contexts

TyAnna Herrington

It is not news that online media have shrunk the world and made it possible for technical communicators living in different nations and regions to collaborate on a range of activities. This kind of global reach allows projects to broaden in terms of who may participate in them and how. This reach has also allowed businesses to expand from local or regional operations to become participants in a growing global marketplace. Such opportunities for collaboration, however, require care in approaching workplace interactions, for with new contexts come new challenges, both anticipated and unexpected. In such situations, the more participants know about the variables affecting international interactions, the better they can anticipate certain problems or respond to others.

In expectation that cultural divergence could arise as a key factor affecting international exchanges, technical communicators need to prepare for potential conflict resulting from such differences. They must also prepare themselves to expect such situations to arise and consider strategies for negotiating the developments that could result from cultural difference. Fortunately, much has been made of this kind of preparation in the published technical communication research over the last decade. And these works have done an effective job of examining how issues of linguistic, operational, and cultural difference can create a variety of challenges that technical communicators might face when working in international venues. Yet within these discussions, the range of valuable publications on global interaction has not been augmented by an extensive treatment of the legal issues that can shape international interactions until now. Such legal issues, however, could be seen as one of the remaining barriers to affect the sharing of ideas and information in the modern global village.

In essence, the law's function is to dispense with conflict by regulating interactions between or among parties who are bound together through place or activity. The law's function is also to enable policy decisions. These combined features make the law a conduit of power used as a source of control. Moreover, in a world where online media have caused traditional barriers of time and distance to almost evaporate, it is the legal boundaries created by different nation-states that seem to impose the few remaining limitations on how information is exchanged in online global contexts. For these reasons, both academic and workplace technical communicators can greatly benefit from a resource that can provide a foundation for understanding topic areas in which legal factors play a pivotal role. *Legal Issues in Global Contexts* is just such a work, for it greatly enriches the work done in technical communication by contributing an important overview of a number of legal issues that affect technical communicators' work in international venues, particularly those created via online media.

To appreciate the importance of how legal issues affect communication in global contexts, one must first understand how the influence of power in legal interactions is even more pointed in international than it is in domestic venues. Domestic law is tempered by the immediate influence of the society in which it exists. It is a context in which the effects of regulation and of finalized decisions are directly perceivable by the population governed by such laws. Moreover, law reflecting public interest is often established on a foundation of an overriding aspirational mission. This mission helps ensure that those who apply the law adhere to the policies that are its purpose. If not, the citizens of the related society can generally turn to various legal or social mechanisms to address perceived misuses of the law. This tempered form of law is generated and employed for the greater good as a means to lead to positive outcomes for the public. It is a situation in which those who have power to influence its contours also help to shape its impact on those it touches.

International law, by contrast, is most often mediated among parties with economic interest as a central factor for negotiation. In such contexts, the greater public is generally removed from the crafting of legal practices and the considerations of legal policies. The general public, for example, is rarely (if ever) involved in the negotiation of treaties for international commerce. As a result, societal interest is not always considered in reaching a final compromise in such international negotiations. For instance, as Potts points out in her chapter in this collection, the inclusion of digital rights management (DRM) tools has significant implications for how we access information and on what we are able to do with that information once we have it. In fact, the very notion that DRM tools are embedded in the books, games, films, and other forms of creative expressions that embrace cultures is significant as an example of information control in broader global contexts. In this case, those individuals with little power have no recourse to reject an international legal decision that affects them, and cultural construction at international borders is not fluid nor is it open.

Even in discussions about how to compromise in international treaties or administer international regulations to provide for the needs of all parties, finding means for such compromise is complex and difficult. Legal scholars, for instance, have argued in favor of extending the fair use doctrine in intellectual property to international intellectual property treaties. Scholars who support this change want treaties to provide public access to otherwise restricted materials as a means to support academic criticism, educational use, free speech, and other needs that allow for transactional and transnational interaction. Such ideas are founded upon principles in the U.S. Constitution and this legal perspective of access and information sharing makes fair use a powerful tool for supporting democracy and enabling egalitarian interaction.

The legal concept of fair use, however, is neither enabled by nor functional within international law because international law is not based on a constitutional structure to ensure its operation. In fact, a strong access-oriented policy in international treaties might even lead to more harm than good for underrepresented entities in treaty negotiation. For instance, emerging countries' cultural assets in art, music, or herb-based medicine can also be their main sources for active interaction in the global community. The ability for such nations to protect these cultural assets from being usurped by more powerful global actors could be hindered by a fair use policy that fails to protect those whose bargaining power is weak. In response to such situations, I have argued that using a model based on a balance of power to allow international negotiation in consideration of humanistic goals is a more reasonable approach, regardless of how complex it would be. Yet to make this approach operate, those who ask for this kind of change would have to understand the intricacy of the legal issues that arise from related discussions.

Understanding the impact of legal issues in international technical communication requires being exposed to the intersection of the law, international venues, and the field. *Legal Issues in Global Contexts* provides a crucial first step for technical communicators by providing a range of vantage points from which to explore legal issues in international technical communication. And it does so by examining how such legal issues affect, or are affected by, various online media.

Now that technology enables the world's population to interact socially, economically, and politically, we must consider how the law that directs our actions in international venues might shape not only future interaction in international cyberspace but also cultural development and information access. In considering such issues, technical communicators need to remember that international law controls our ability to create and be supported by our innovations in international venues in the same way that domestic law controls our actions at home. If we want to participate in the world as global citizens, we will also have to consider how our participation will be regulated, guided, and in some cases, restricted. We might derive practical benefits through our interactions in global markets, but we might also derive the more abstract benefits of cultural and political participation in the world's interactions. Issues of free and open speech, authorship, privacy, and information access form the basis of complex difficulties in domestic settings. These same issues can lead to even more complex questions in international venues.

For instance, Kraglund-Gauthier and Young's chapter notes that we need to be aware that once we use digital media, our privacy is less protected. By extension, of course, we should be aware of the fact that the digital media sending information to the world affects our privacy much more broadly than it would in local spaces. An idea that might form the basis of what seems to be local discussion can, when considered through interaction in digital form, be broadcast worldwide and engender a much broader global debate. But what might happen if, for instance, a student grappling with new concepts expressed ideas that might seem extreme to some of the participants in a discussion? In the more local, even if digital, learning venue in which the context of discussion would be well formed and the cultural constructs for discussion would be well established, the student's comments could be considered as a part of the process of exploration. In this kind of situation, a student might safely "test" the limits of speech. In so doing, the student might make potentially inflammatory statements in the heat of discussion as a way to test ideas. In the "private" venue of a classroom, the consequences of such testing would be limited. If, however, privacy were inhibited and the student's statements were broadcast globally, the impact of such statements could be extreme. On the one hand, such public contexts might make the student a target for retribution. On the other, these settings might create a false impression of the power of the statement as representative of a culture as a whole. And as Kraglund-Gauthier and Young point out, information can be relatively permanent when put online. Thus, statements that might have been pertinent and potentially even appropriate in a specific local context might lead to long-term negative effects in global venues.

Moreover, when we consider issues surrounding privacy in global contexts, we should also be aware that related communication concerns are not affected only by distance. Rather, cultural input and influence can greatly determine what might or might not be acceptable as a part of a global conversation. Controversies surrounding depictions of Muhammad are illustrative of such influences. To some, all depictions of Muhammad are sacrilege; to others, they are a matter of protected speech. In both cases, the culture of the individual markedly influences what is considered acceptable or not. Additionally, as Lunceford points out in his chapter, the emotional nature of our responses to issues such as these is made more complex by international online access that requires us to become aware of the rhetorical nature of the language we use when discussing ideas.

And the rhetorical context in international venues would have to be determined within a range of diverse cultural constructions.

Such concepts are made further complex when they take place in virtual worlds that represent alternate or parallel realities. Chávez-Aguayo, for example, writes about *Second Life* and considers how virtual and real worlds interact. I conjecture that globally accessible, virtual worlds such as Habbo Hotel, RuneScape, and World of Warcraft are precursors to new virtual worlds that could someday become a kind of "third places." These third places would be online spaces that help diffuse cultural difference by amalgamating multiple cultural characteristics as a basis for their formation and participant interaction. However promising a culturally diverse virtual world might be, legal questions would likely arise with regard to who would control these spaces and which country's laws would apply when and where. And in a global online context, who would provide or inhibit entry, what kinds of contractual arrangements would allow or inhibit interaction, who would control the intellectual products created, and how would they be controlled? Thus, the process of developing and participating in a culturally nonspecific global virtual world could quickly generate international legal questions that individuals need to consider in advance.

Other less abstract but no less complex questions include those observed by Batova, who points to problems with translation when sharing information across cultures. To underscore her call for improved legal literacy and more exact translation of terms in the law, I ask that we also consider another looming issue in legal literacy—the difficulty in finding means to translate connotations of legal terms when the conceptual structure behind them still is in stages of argumentation. A relatively mundane example is in the connotations of the word *monument* in international contexts. In an instance of collaborative interchange between American and Russian students in the Global Classroom Project, students decided to study cultural differences between monuments in St. Petersburg, Russia, and those in Atlanta, Georgia. During this project, the students experienced great difficulty in agreeing upon the subject matter for study, and they failed to understand where communications broke down. As it turned out, the Russian connotation of МОНУМеНТ is less inclusive than the English *monument.* Once the students understood where they differed in connotative understanding, they soon agreed to a common subject of study. Legal disagreements can likewise arise when terminology leads to a seeming agreement, but the parties later realize that their bases for agreement were on differing assumptions. These issues at the intersection of law, international interaction, and technical communication will thus become more critical as our global business interactions, educational efforts, and social exchanges continue to broaden.

Regardless of local or global context, there is an ongoing relationship between technical communicators and their community. These relationships require technical communicators to continually consider their roles as conduits of information, and such considerations affect what technical communicators do in different settings. For example, Copenhaver, Vee, as well as Zemliansky and Zimmerman examine different aspects of intellectual property within a broader global context. Vee points to the limiting effects of patent law on educational software development, and Zemliansky and Zimmerman discuss how copyright law has changed approaches to developing new texts via digitization. Copenhaver examines legal developments in consideration of orphaned works and emphasizes the need for public access as a means for participating in the global community. In so doing, these authors, both individually and collectively, create a framework for understanding such issues.

The various topics and perspectives presented in the chapters of this collection create an effective collective mechanism for supporting discussion of power as a necessary accessory to participation in global legal negotiations. These collected chapters also provide a framework for understanding the various factors that can affect who will have a voice—or who might acquire a voice—when it comes time to debate legal issues that might affect the greater global community. By examining these issues, this text emphasizes a key central point: Those who work in global realms or cross physical borders through digital access are no longer merely citizens of their nations; rather, they have become citizens of the world. As such, they can influence the shape of the world in which they will interact. Yet they can do so only if they have power to introduce innovations to the world market, to affect negotiations for the rules of border crossing, and speak about choices that influence the lifestyles and livelihoods of the world community. As the different entries in this collection emphasize, the influence of legal choices can be significant. Thus, for technical communicators working in such contexts, an understanding of such legal issues has become essential to not only their success as professionals but also to their participation as citizens in a greater global community.

In closing, I suggest that if technical communicators have the power to shape the global community through their translations, business interactions, teaching, and writing, then they should do so with humanistic interests as their primary concern. Instead of focusing on what it takes to deal with cross-cultural conflict in crossing global borders, we might consider the vantage point of finding common desires for humanistic goals and develop cooperative efforts on the basis that humanistic needs, desires, and driving influences are commonly shared among global entities. If international law were established primarily on the basis of human needs and secondarily on the basis of market needs, then market needs would serve the people of the global community rather than in reverse. Addressing these various issues is no easy task, but the entries in this collection provide the reader with a framework for understating the central issues involved in such processes and provide readers with a mechanism for making informed and effective decisions for action in global contexts.

INTRODUCTION

Legal Issues in Global Contexts: Examining Friction Points on the Flat Earth

Kirk St.Amant and Martine Courant Rife

It has been said that we live in the era of globalization. If that is the case, then the slogan for that era is undoubtedly "The world is flat." And with each passing day, we are reminded of how new technologies are leveling the old barriers of time and distance to open a newer, flatter geography of global communication. It is a context in which individuals located on different continents can collaborate on project teams. It is a context in which music recorded in New York can be accessed in New Delhi within seconds of its creation. It is a context in which the very word "barrier" seems on the verge of becoming an anachronism. Or is it?

Oddly, as real barriers of physical space have been flattened by the human construct of cyberspace, another human-made barrier has emerged to limit interactions in that space. That new barrier? Law. In this new context, just because one *can* send information instantaneously across the globe does not mean the intended recipient will be *allowed* to access or to use that information.

These legal impediments force us to reconsider the idea of the flat earth as not a uniformly smooth landscape. Rather, it is a surface covered by *friction points*—areas in which something can slow, or even stop, the flow of ideas and information. Thus, exchanging information over the flat earth is akin to sliding a disk across the surface of a frozen pond. One needs to know where the friction points are in order to avoid them and get the object from point A to point B as quickly and easily as possible. This situation is particularly the case in technical communication. As the field is based on the exchange of information, these friction points will markedly affect how technical communicators participate in the modern global economy. And in the end, it all comes down to one central issue: content.

To begin, the friction points that impede the global flow of online information are not random. Rather, they often reflect deep-seated cultural attitudes concerning appropriate behavior. National laws, in turn, represent the codification of a particular culture's or group's attitude concerning what is and what is not acceptable content to create, access, or exchange. In the age of globalization, these legal differences can create barriers as real and as effective as any fence or wall. In the United States, for example, national law allows citizens to access and exchange a wide range of content, including critiques of the government or of public figures. In other nations, however, laws might greatly restrict or prohibit the creation of or access to such content. As a result, variations in national laws can affect how easily and effectively certain content can move from point to point across the flat earth. This factor has profound implications for technical communicators working in global contexts.

Increasingly, many of the international legal differences that generate friction involve the creation and the management of content. Copyright law, for example, is one of the greatest friction points affecting the free and open flow of ideas and information across the flat earth, for it stipulates who has the right to replicate or to share certain kinds of content. Similarly, perspectives on privacy rights—and who can legally access and copy an individual's personal information (i.e., content)—can differ from nation-state to nation-state. As a result, when individuals try to share different kinds of content/information across borders, legal systems can intervene to create friction that slows or stops that exchange.

Different national laws can also determine the language in which content can be provided or exchanged in different geopolitical regions. The European Union (EU), for example, requires that much of the content distributed across its member nations be translated into the languages of those member states (Language requirements, 2012; Walmer, 1999). Such legislation, moreover, can affect the creation and distribution of content within nations themselves. To distribute information/content in the Canadian province of Quebec, for example, organization are often legally required to provide that content in the province's official language of French (Medical device regulations, 1998). More recently, demographic shifts have created new legal contexts affecting the language in which technical information (content) must be provided to individuals within a specific nation, as is currently the case of providing medical-related information within the United States (St. Germaine, 2010). All of these factors have important implications for technical communicators whose primary job is to create and manage content. For these reasons, today's technical communicators need to understand these legal friction points if they wish to work effectively in global contexts.

To address these global friction points, technical communicators must combine their knowledge of content creation and content management with an understanding of the international legal issues affecting those practices. The objective of this collection is to provide an overview of the different legal aspects that can be friction points one might encounter when working with content in international environments. Through examining these topics, this collection provides readers with an initial and a relatively broad-based review of legal issues affecting technical communication practices in global contexts.

The entries in the book's first section, "Challenges and Considerations," provide an overview of the new and unique situations created by extending legal concepts and practices into global contexts. In this section's first entry, Wendy L. Kraglund-Gauthier and David C. Young examine how technical communicators should rethink the notion of privacy—both what it is and how it is protected—when interacting in international situations. To do so, the authors use a case study that examines the often misunderstood issue of privacy laws in online education. Through this case study, Kraglund-Gauthier and Young discuss how individuals and institutions might rethink the idea of protecting the rights of online users. They also make recommendations for how such issues could be addressed within international online interactions.

In the section's second entry, Brett Lunceford reviews the debate surrounding network neutrality, a situation in which technical details have been largely overlooked in favor of slogans and arguments that appeal to heuristics. Lunceford notes that, despite the complexity of the issues surrounding Internet traffic in international contexts, proponents of network neutrality have centered the debate on the master term of *neutrality*. He also examines how the rhetoric surrounding net neutrality can provide technical communicators with insights into how technology issues are rhetorically constructed in the public sphere. In so doing, he notes how such constructions affect international law and public policy and overviews the difficulty of governing a global network that transcends national borders and jurisdictions.

The section's third entry, "The Rules of the Game: Real Legal and Economic Implications of *Second Life*," takes the reader even deeper into the intersection of online media and global interactions by examining the online virtual world of *Second Life*. As the author Marco Antonio Chávez-Aguayo explains, these contemporary computer technologies now connect artists and producers with audiences around the world. However, such environments also pose new challenges for technical communicators who must grapple with legal issues such as copyright, privacy, and the distribution of private information when using *Second Life* to participate in internationally distributed teams.

Ronald L. Stone's chapter, the final entry in this section, moves the discussion of representation in global contexts further toward the abstract as he examines how numeric representation has legal implications in international interactions. Specifically, Stone examines the legal difficulties created by using different systems of measurement in global contexts. He also examines how metrication—the use of the metric system—creates legal communication challenges in international settings. In his analysis of these issues, Stone examines common problems technical communicators might experience in such situation and overviews prospective solutions to such problems.

Entries in the book's second section, "Language and Access," examine the legal challenges technical communicators face when trying to create information that can be used effectively in global settings. This section begins with Tatiana Batova's discussion of the idea of legal literacy and its connections to both technical communication and technical translation processes in global contexts. Specifically, Batova looks at how multilingual technical communication projects require a new emphasis on civic engagement in technical communication in order to better understand the legal contexts in which technical translation occurs. In so doing, she provides a framework for understanding the complex area in which law, technical communication, and technical translation intersect.

In the section's second entry, Nicole St. Germaine continues this examination of legal issues and language by reviewing how technical communicators are affected by Executive Order 13166. St. Germaine does so by presenting both the results of her research on the literature associated with this Executive Order and her interviews with technical writers and health communicators who are currently writing technical documentation for different cultural audiences. She then concludes her chapter by presenting recommendations—or best practices—for how technical communicators might address specific issues relating to translation and information sharing in cross-cultural contexts.

Usability, from a global perspective, often involves more than just language. It also involves access to information. Such access, however, is often regulated by a patchwork of different national laws that affect who can access certain kinds of information when and where. Liza Potts examines this notion of access in global contexts by reviewing the concept of digital rights management and associated practices in global contexts. As Potts explains, while individuals around the world can use online media to access digital entertainment through various sources, locating, watching, and interacting with this content is riddled with usability issues that can create a deterrent to legal use. In her chapter, Potts traces these issues in order to reveal the different usability concerns involved in these situations. She then concludes by calling for improvements of the user experience and of industry policies regarding digital rights management and the Digital Millennium Copyright Act.

Online media evolve continually and rapidly, but legal change is often slow and sporadic. As a result, the online environment seems to continually be one, if not one hundred, steps ahead of the legislation that tries to regulate it. The book's third and final section, "Ownership and Authorship," explores some of the emerging issues that can contribute to confusion with regard to law in global contexts. As Pavel Zemliansky and Traci A. Zimmerman explain in the section's first entry, recent developments in technical communication theory and practice have yet to address the importance of understanding intellectual property and authorship. In order to better comprehend the implications of this gap, Zemliansky and Zimmerman review major developments in copyright and intellectual property affecting the education of today's technical communicators. They also examine how that education manifests itself in professional training and professional practice as well as provide suggestions for engaging effectively in such practices.

Annette Vee continues this discussion of intellectual property in global contexts by analyzing the effects of patent law in writing studies. In so doing, she reviews how the monopoly rights granted through patents can affect composition in, of, and with software. She also outlines some specific reasons patent law pertains to writing specialists, and explores the blurring boundaries between written text and computer code. Vee then concludes with an examination of controversies surrounding patent protection for software in international contexts.

The section's final entry combines ideas of copyright and access to examine the relatively overlooked area of orphan works. In this chapter, Martin S. Copenhaver overviews the legal concept of orphan works, or copyrighted works for which the copyright holder cannot be identified. Through this overview, Copenhaver discusses the implications such works have for accessing and using information in an age of the global Internet. He also reveals why such works should be seen as important to international technical communication practices and uses theoretical and practical considerations to explain the global legal and policy significance of orphan works. Copenhaver then concludes by providing technical communicators with suggestions for how they can participate in the ongoing global dialogue on this important legal issue.

Through examining these topics, the various chapters in this collection provide readers with an initial and a relatively broad-based review of legal issues affecting technical communication practices in global contexts. The editors hope readers will use the ideas and information presented here as a foundation for creating best practices and engaging in future research in the area of legal issues in international settings. By working together to examine such issues, technical communicators can create methods for addressing friction points and improving the flow of content across the newer, flatter earth.

REFERENCES

Language requirements in Europe. (2012). *Net-translators*. Retrieved from http://www.net-translators.com/knowledge-center/articles/language-requirements-in-europe

Medical devices regulations, SOR/98-282. (1998). *CanLII*. Retrieved from http://www.canlii.org/en/ca/laws/regu/sor-98-282/latest/sor-98-282.html

St. Germaine, N. (2010). Technical communication in the health fields: Executive Order 13166 and its impact on translation and localization. *Technical Communication, 57,* 251–265.

Walmer, D. (1999). One company's efforts to improve translation and localization. *Technical Communication, 46,* 230–237.

PART I

Challenges and Considerations

http://dx.doi.org/10.2190/LIIC1

CHAPTER 1

Hiding Behind a Password: Are Online Classes as Private as We Think?

Wendy L. Kraglund-Gauthier and David C. Young

Since the proliferation of the Internet as a medium of information, we have witnessed a veritable explosion in online opportunities for learning. To address student demand for accessible course offerings, institutions of higher education have expanded their methods of course delivery from the traditional brick-and-mortar classroom to now include "bricks and clicks" and just "clicks." As Grant-Davie and Cargile Cook (2005) have noted, technical communication program administrators and designers have been incorporating digital media into content design as quickly as technology has been evolving. However, this trend toward ever-increasing availability of online instruction does pose problems for all stakeholders.

We contend that domestic privacy laws such as Canada's *Personal Information Protection and Electronic Documents Act* (PIPEDA), as well as parallel legislation that exists in the broader international community, lacks specificity and stringency. In fact, there is a lack of clarity in terms of protection surrounding the collection and storage of course-based materials, including online discussion board content and class recordings. With the legitimate concern of many individuals that their personal privacy be safeguarded, have we established a false sense of security among online users when we offer password-protected sign-ins? Furthermore, because of the cross-border nature of online learning, there is a certain amount of ambiguity regarding the application of regulations and laws.

Failure to acknowledge this new reality created by online education in its various forms can pose potential ramifications for how we communicate in a global and technical world. In this chapter, we explore the underresearched, underacknowledged, and often misunderstood issue of privacy in terms of electronic learning as a medium of education. We discuss how individuals and institutions of higher education should be more attuned and committed to protecting the legal rights of online users. We also offer recommendations and alternatives for consideration. Although privacy is expected, it should not be assumed.

DEFINING ONLINE EDUCATION

> Learning is a continuous, cultural process—not simply a series of events. It values and transcends the classroom and the workplace. Access and opportunities to learn should be available to anyone, anywhere, and at any time.
>
> —Rosenberg, 2001, p. xvii

The Canadian Council on Learning (2009) has defined electronic learning (or e-learning) as the development of knowledge and skills through the use of information and communication technologies; in particular, e-learning is "the application of computer technologies to education" (p. 4). Other terms include web-based learning, digital learning, and virtual learning. Simply put, e-learning refers to learning that is facilitated by the Internet and by computer technology; collectively, this technology is referred to as information and communications technology (ICT).

Synchronous and Asynchronous Learning

Online learning options are characterized as being either synchronous or asynchronous. Synchronous online learning most closely resembles traditional face-to-face learning in which physical and temporal space merge. Technological advances have led to myriad collaborative tools and learning environments; hardware and software are now relatively inexpensive and user-friendly. Microphones are used to transmit voice in real time, and live visual images are shared via video capture devices, or webcams. Proprietary online platforms that afford synchronous possibilities include Elluminate Live, Wimba, and Second Life.

These platforms share similar characteristics: a roster on which the names of the students in attendance are displayed, a chat feature in which participants can send immediate messages privately to other participants or publically to the entire room, and access to session recordings. As in the traditional face-to-face class setting, students gather at a prearranged time at a prearranged location for the specific purpose of learning course materials. Yet in the case of a virtual classroom, that prearranged location is at each participant's private computer workstation or whichever Internet location they so choose.

In comparison, asynchronous learning is not constrained by time. Using programs such as Blackboard, WebCT, Moodle, Desire2Learn, or institutionally developed platforms, just to name a few, enrolled students log into their classrooms and complete assignments, read and respond to peers' postings on discussion boards, and download learning materials. All this can be done at the students' own time, regardless of time zone. Many asynchronous classroom options also allow for online testing, assignment uploads direct to the course facilitator, and repositories of class materials and external links.

Despite a lack of immediacy afforded by asynchronous platforms, students and professors in asynchronous classrooms report feeling connected with each other via their internal postings. As well, in our experience, students and professors have also noted the idea that the asynchronous nature of the classroom has actually led to students making more in-depth reflections and displaying critical thinking through their postings. They have reported that the asynchronous nature of the course has meant they have had time to reflect on their readings, read their colleagues' postings on the same topic, and then carefully formulate their own response.

Whether working in synchronous or asynchronous learning environments, when developing a graduate level course to be delivered via online modalities, content creators should have an awareness of domestic and international privacy laws. Often, course content is delivered and stored in electronic form. These recordings contain text, audio, and visual components of class sessions. As instructional designers and professors, we have both witnessed instances wherein students have shared information in an online setting. In an alternative context, this activity could be construed as divulging information that may be considered a violation of privacy laws. Although established guidelines exist concerning confidentiality, in fact, students in dynamic debate with colleagues during online class sessions may reveal factors that would specifically identify people and places. For example, casually referencing a particular student, teacher, school, or school board may occur as an innocent comment made during conversation. In face-to-face exchange, words can be ethereal; in an online context, that moment in time is captured and stored—often in perpetuity.

THE PROLIFERATION OF ONLINE LEARNING

> Learning is key to productivity, competitiveness and prosperity. E-learning blends the technology that pervades our work with the constant skills-upgrading needed to work with technology.
>
> —Bloom & Murray, 2001, p. 38

The pervasiveness and power of the Internet is often difficult to conceptualize, yet when placed in the context of radio and television, its potential reach is astounding. In little more than three years, the Internet gained an audience of over 50 million users; it took television 15 years and radio 37 years to achieve the same audience-base (Naughton, as cited in Curran, 2008). Over the past 10 years alone, technological advances have significantly shaped the way individuals communicate, build relationships, shop, and learn; the veritable explosion of ICT has vastly changed the educational virtual landscape.

Why Online Learning?

John Chambers, Chief Executive Officer for Cisco Systems, a leading provider of network software and hardware, has purported that "the biggest growth in the Internet, and the area that will prove to be one of the biggest agents of change, will be in e-learning" (as cited in Rosenberg, 2001, p. xiv). For varied reasons, including providing accessible programming, addressing student demand, and responding to competitive and economic pressures (Abrami et al., 2006; Canadian Council on Learning, 2009; Naidu, 2006), more and more students are self-selecting online modalities to initiate or further their learning.

Rosenberg (2001) has highlighted 11 key benefits to e-learning that encompass the needs of both students and organizations. These benefits are

1. Lower costs for learning
2. Enhanced business responsiveness
3. Consistent or customized messages, depending on the situation
4. Content is more timely and dependable
5. Learning is 24/7
6. Little user "ramp-up" time
7. Employs a universal platform
8. Builds community
9. Scalability
10. Leverages the corporate investment in the Web
11. Provides an increasingly valuable customer service (pp. 29–31).

The assortment of online course options means students have increased opportunities to select courses that meet individual needs, requirements, and interest and skill levels. Students' options of programs that feature continuous enrollment options, self-paced modules, and flexible scheduling translate into more opportunities. Students are able to schedule classwork around personal and professional responsibilities, and with access to a computer and the Internet, they can study anytime and anywhere. Varying online course delivery methods that incorporate individual and group work, text-based, audio, visual, and other multimedia formats also address differing learning styles. As further incentive, some institutions promote minimal program admission requirements and the transfer of existing course credits from other institutions in an effort to capture the lucrative online market.

Online Learning as a Training Medium

Technical communication skills have become increasingly important in today's economy. The emergence of transglobal enterprise and multinational corporate offices connected via digital means has created a need for the workforce to be well versed in technical communications. Employees can often learn new content or upgrade existing skills via online modalities. St.Amant (2007) has noted the increased international interest in technical communications—specifically how "some employers believe that the experience of interacting within multinational groups in online classes adds value by teaching students how to communicate within a global context that mirrors modern business practices" (p. 16).

Online learning technologies are often touted as a panacea for addressing learning needs in challenging economic climates. The potential cost savings in travel, accommodations, replacement workers, and other expenses associated with training can be significant once the initial costs associated with necessary ICT infrastructure have been addressed. Instead, some companies have recognized the greater return on investment in training that is specifically tailored to individual learner's needs (O'Leonard, 2009). Ultimately, online learning programs are a convenient and oftentimes cost-effective way to ensure individuals receive the appropriate and timely training they require.

Growth Trends in Online Learning Opportunities

St.Amant (2007) has linked the growth of international online education with technical deregulation and global interest in learning opportunities. Educational institutions and businesses are becoming increasingly more acceptant of the validity and flexibility of online learning. The Sloan Consortium (Sloan-C) is an American-based, nonprofit organization that focuses on research and application of "online education into the mainstream of higher education, [thus] helping institutions and individuals improve the quality, scale, and breadth of online education" (Sloan-C, 2008, Introduction section). Sloan-C surveyed American universities to gauge online course adoption and reported that by 2002, over 80% of surveyed universities offered at least one online course. By 2006, over 3.5 million students had enrolled in a course offered via online modalities (as cited in Dell, Hobbs, & Miller, 2008). Between the fall of 2004 and the fall of 2005, the enrollment of online students increased 35%, or 850,000 additional students, and there are no indications that an e-learning enrollment plateau has been reached (Ruth, Sammons, & Poulin, 2007). The results of a survey done by the Instructional Technology Council (2008), a consortium of American Community and Junior Colleges based in Washington, DC, indicate that distance learning enrollments increased by 11.3% at a time when enrollment in campuses overall increased by less than 2%, with evidence that the demand for distance learning greatly exceeded the available course offerings. Moreover, the Council's report notes that

> Many colleges had significantly increased their number of blended or hybrid and/or Web-enhanced or Web-assisted courses. Most colleges have rapidly expanded their student services and technology support services to meet accreditation expectations of "equivalency" with traditional face-to-face courses.

In 2007, Statistics Canada (2009) conducted a survey of Internet use across Canada. From a sample size of 26,588 (a 75.9% response rate), they declared that the number of people who reported using the Internet for education, training, or schoolwork increased from 26.1% in 2005 to 34.0% in 2007 (+7.9%). This percentage increase was surpassed only by the number of respondents who reported obtaining and saving music (+8.2%) and using an instant message program (+11.2%). In another study conducted in 2005, just over one-quarter (26%) of adult Canadians—an estimated 6.4 million—reported they went online for distance education, self-directed learning, or correspondence courses (McKeown & Underhill, 2005). While residents from rural and small town areas were less likely than their urban counterparts to report going online to research assignments, they were more likely to report going online for educational purposes.

Similar statistics have emerged from Europe. As Curran (2008) has reported, the prevalence of ICT has continued to grow steadily since the arrival of the new millennium. The European Centre for the Development of Vocational Training (ECDVT, 2001) reported that by 2000, e-learning initiatives were embedded into European Union (EU) policy objectives for economic growth and employment strategies.

Specifically, in 2001, online learning accounted for 14% of training expenditures. Curran (2008) also reported on a 2005 study of adults in ten EU member states, in which "12% of the adult population [surveyed] makes use of the Internet as purposeful organized learning" (p. 30). However, online course enrollment in the EU is not as popular; a mere 8% of students enrolled in full-time courses report taking an e-learning course.

In developing countries, distance education can serve as a viable alternative to the steep capital expenditures of brick-and-mortar classrooms. On revenue-starved campuses, establishing strategic business alliances and partnerships with external academic institutions that offer online site hosting and course delivery is one cost-effective solution to increasing enrollments. Weigel (2002) has argued that large-scale introductory courses such as English literature or biology, which tend to have similar content from university to university, are prime targets for the corporatization of online learning. The Internet itself has served to ease the transfer of credits between institutions. Course materials are often posted online and verifiable with a click of a mouse. Examinations can be sent to proctors via email, and students can be tested via online test banks.

Geographical space is shrinking under the practically limitless reach of the Internet; a virtual classroom can easily be filled with students from all over the globe. As Abrami et al. (2006) argue, an online learning presence requires universities to have policies and strategies to meet the emerging student and faculty demand. However, some authors have suggested that Canada lags behind other countries in their research in and adoption of e-learning policy and strategy (see Abrami et al., 2006; Canadian Council on Learning, 2009; Charpentier, Lafrance, & Paquette, 2006). In their "Vision Statement on On-Line Learning in Postsecondary Education," the Council of Ministers of Education, Canada (CMEC, 2001) urge governments and postsecondary institutions to hasten and coordinate their efforts to increase online course offerings. They note the power of online learning to narrow the digital distance between Canadians and recommend processes that promote "quality assurance, transferability, and high technical standards" (p. 85).

Online education is purported to help reduce potential barriers to accessing education, such as cost and distance. While more definitive research on this matter is required, this does suggest that the Internet and its infrastructure can be harnessed as a tool to help overcome the barrier of distance in education (McKeown & Underhill, 2005). In summary, there has been a substantial increase in the offering of online courses at both American postsecondary and public schools in the past five years. It is not unreasonable to suggest that similar trends are reflected in Canadian educational circles. As online learning gains momentum globally, its increase in popularity aligns with technological advances in hardware and software. The technology that is used has improved dramatically, resulting in an increased demand by students for this form of program. It is incumbent upon institutions to clarify their intentions around online learning and to ensure that learning platforms and their support systems for students and faculty be of the highest quality.

ONLINE EDUCATION AND THE LAW

> Information technology has the capacity, if not to end privacy, then to redefine what we mean by the term.
>
> —Boyle, 1996, pp. 3–4

Privacy Law: Setting the Stage

Writing in 1890 in the *Harvard Law Review*, Warren and Brandeis noted in their famous piece "The Right to Privacy" the legal and constitutional "right to be let alone" (p. 193). In the more than 120 years that have elapsed since then, privacy has become an ever-increasing ethical and legal issue. Arguably, the concern over privacy has been exacerbated by the seemingly unending proliferation of technology. Although most modern democracies have enacted legislation that purports to protect personal privacy, there are countless examples of violations of this privacy. In fact, oftentimes it is the very agencies entrusted with maintaining privacy that breach these established safeguards. For instance, the endless tales of wiretapping and surveillance carried out by the Federal Bureau of Investigation during the directorship of J. Edgar Hoover serve as a stark reminder that privacy is not unassailable. Additionally, reports of increased government encroachment into the realm of personal privacy in the aftermath of the September 11, 2001 terrorist attacks have become seemingly commonplace. Thus, it would seem that although personal privacy is often viewed as sacrosanct, it can and at times is subject to certain limitations or restrictions.

What is Privacy?

An issue that remains somewhat perplexing for many writers and commentators is what the concept of privacy actually entails. Clearly, as Fried wrote in 1968, "The rights of privacy are among those basic entitlements which men must respect in each other" (p. 484). Although Fried's words facilitate the establishment of privacy as some type of terra firma upon which social relations should be constituted, it does little to shed light on defining privacy. In fact, the term is a contested one, and uniformity of thought surrounding its meaning is difficult to establish or pinpoint. As McCullagh (2008) noted, "A single definition that adequately incorporates all the subtle differences that privacy evokes has so far proven impossible" (p. 4). As vexing as this might be, it is wholly inadequate to proceed without offering some statement of meaning for the word privacy, even if this statement is imperfect and incomplete. In this regard, revisiting Fried's work is an appropriate starting point.

For Fried (1968), although there is an inextricable link between privacy and secrecy—the latter, which he characterized as the capacity to restrict knowledge about oneself to others—privacy should not be confused with secrecy. As he pointed out,

> It is not true, for instance, that the less that is known about us the more privacy we have. Privacy is not simply an absence of information about us in the minds of others; rather it is the control we have over information about ourselves. (p. 483)

This concept of personal control is important, as it moves the debate toward a notion of self-autonomy.

In this vein, DeCew (1997) believes that privacy must be seen as a broad cluster concept that involves "our ability to control information about ourselves, our ability to govern access to ourselves, and our ability to make self-expressive autonomous decisions free from intrusion or control by others" (p. 62). Thus, for DeCew, the cluster involves personal control over the following:

- Informational privacy (control over one's information);
- Accessibility privacy (involves physical privacy, in the sense of who is granted access as well as the degree of this access).

Building on the contributions of Fried (1968) and DeCew (1997), it might now be possible to advance a rudimentary working definition of privacy. While privacy may be bound with secrecy, privacy transcends this union, to encapsulate multiple interests that each involves personal control. Thus, privacy becomes, to borrow Gerety's (1977) phrase, an "island of personal autonomy" (p. 271). In sum, privacy is "a shield for us in various contexts" (DeCew, 1997, p. 79).

Privacy: The Legal and Policy Dimension

For this proverbial shield of privacy protection to have any real effect, it cannot operate in isolation. Rather, personal privacy must be buttressed by some external mechanism with widespread application. Thus, legislation has been enacted at the national, state/provincial, and local levels in the name of safeguarding personal privacy. An exhaustive examination of privacy laws is beyond the scope of this chapter. It is, however, beneficial to provide even a cursory examination of some of the major legislation currently in force in various countries, with particular reference to online privacy.

As a backdrop, it is important to recognize that most privacy and data protection laws are structured around the notion of "fair information principles." Although these principles have been interpreted in various forms, most legislation reflects the 1980 recommendations of the Organization of Economic Co-Operation and Development (OECD). According to the OECD (n.d.), personal data collection should be subject to the following guidelines:

- Collection of personal data should be limited in scope, and conducted with respect to legal protocols.
- Any personal data that is collected should be relevant, accurate, as well as up-to-date.
- The purpose for data collection should be specified, and the use of such data restricted to fulfilling this stated purpose.

- Personal data that is collected should not be disclosed, except under certain conditions.
- Reasonable safeguards must be established to protect any personal data that is collected.
- A policy of "openness" surrounding personal data should be established.
- Protocols should be enacted to permit an individual to inquire as to whether personal data has been collected about him or her, and if so, whether the information collected is accurate.
- Finally, those responsible for controlling data should be held accountable for ensuring that these guidelines are respected.

These recommendations, now almost three decades old, continue to have a lasting legacy on current privacy laws, and their imprint is readily discernible in legislation throughout various OECD member states.

Canadian privacy legislation is no exception, and in fact, even a cursory examination of the various polices demonstrates a recognition of the OECD guidelines. On the national stage, Canada has two relevant pieces of legislation: the Privacy Act,[1] and the Personal Information Protection and Electronic Documents Act (PIPEDA).[2] The former, which came into effect in 1983, applies only to federal government institutions and details the manner in which these institutions can collect, use, and disclose personal information. Of greater significance to the current discussion is PIPEDA, which was enacted in three stages between 2001 and 2004. This law, which relates to data privacy, applies to federally regulated institutions as well as private sector organizations. Essentially, PIPEDA governs how private sector organizations collect, use, and disclose personal information in the course of business transactions (Minister of Justice, 2009). Thus, PIPEDA "promotes and enforces a unified privacy principle across Canada" (Abramowicz, 2006, p. 57).

However, besides establishing a legislative and legal framework concerning online privacy, PIPEDA also permitted Canadian compliance with the European Union's 1995 Data Protection Directive. This act, which harmonized data protection legislation among the member states of the European Union (Information Commissioner's Office, n.d.), is built on eight core principles that stipulate that personal information must be

1. Fairly and lawfully processed
2. Processed for limited purposes

[1] It should be noted that all Canadian provinces have enacted provincial public sector privacy legislation.

[2] British Columbia, Alberta, and Quebec have enacted provincial private sector privacy legislation that supersedes PIPEDA.

3. Adequate, relevant, and not excessive
4. Accurate and up-to-date
5. Not kept for longer than is necessary
6. Processed in line with your rights
7. Secure
8. Not transferred to other countries without adequate protection

In stark contrast to the Canadian and European Union's approach to data privacy is the American position. Relying on what has been termed a "sectoral" approach to data protection, the United States has yet to enact a national privacy law comparable to Canada or the EU. In fact, it is interesting to note that few American states even recognize an individual's right to privacy.[3]

Thus, what remains is a hodgepodge of various pieces of legislation, often applicable to only certain sectors, and much of it adopted when situations have warranted. Despite the lack of an American policy, the United States, much like Canada, has reached compliance with the European Union's *Data Protection Directive*. The *Safe Harbor Principles*, which apply to companies in the United States and the EU that store personal data, are designed to prevent accidental disclosure or loss of information. In order to join, an organization must comply with the seven safe harbor principles of notice, choice, onward transfer, security, data integrity, access, and enforcement (export-gov, n.d.).[4] Laws regulating online privacy remain problematic; as more countries enter into the online learning milieu, the lack of clarity surrounding domestic and international privacy laws will inevitably lead to the emergence of future issues in the management of online learning programs.

ONLINE EDUCATION AND PRIVACY LAWS: IMPLICATIONS AND FUTURE DIRECTIONS

> Any social interaction—any attempt at connection—involves making ourselves vulnerable.
>
> —Palloff & Pratt, 1999, p. 42

When teaching undergraduate and graduate classes in education law, privacy is a topic that often occupies a central position in the course. As professionals, we should be concerned with ensuring that we safeguard any and all personal information with which we deal; as outlined in the previous section, countries have developed various laws, policies, and protocols under the auspices of

[3] It should be noted that California does recognize a right to personal privacy and even enacted the *Online Privacy Protection Act* in 2003, which deals with data privacy.

[4] A U.S.-Swiss Safe Harbor Framework is also in place.

personal privacy. As well, for the vast majority of the students we deal with, privacy appears to be an issue of major concern. It is interesting to note that these same students also indicate that current legislation enacted by various levels of government to protect personal privacy does little to allay their unease.

Public Perceptions of Privacy

In their survey conducted in 2007, Statistics Canada (2008) asked specific questions related to online security and privacy. These questions related to Internet exploration, banking and other financial transactions, and accessing and supplying government-held personal information. As demonstrated in Table 1 (Statistics Canada, 2009), over a 2-year period, the level of concern about privacy on the Internet has remained fairly stable; by 2007, some 34.2% of all respondents reported being very concerned about Internet privacy compared to 36.6% in 2005.

These Canadian results are hardly atypical; a 2008 *Eurobarometer* survey on privacy revealed that 82% of European Internet users have a low degree of trust in online personal data management. Survey results also indicated that more than half of those surveyed felt legislation was incapable of dealing with the increased amount of personal information exchanged on the Internet (European Digital Rights, 2008). Thus, as "many scholars have noted . . . the existing regulations aimed at addressing privacy concerns in the context of electronic communications are inadequate" (Woo, 2006, p. 960).

Table 1. Internet Use by Individuals, by Level of Internet Privacy Concern and Age

	Not at all concerned		Concerned		Very concerned	
	2005	2007	2005	2007	2005	2007
	% of Internet users at home[a]					
Total	24.3	25.4	37.5	38.8	36.6	34.2
34 years and under	28.6	31.5	42.4	42.3	27.0	24.5
35 to 54 years	21.4	21.0	35.5	37.4	41.6	40.3
55 to 54 years	21.6	21.5	31.8	33.0	45.5	43.6
65 years and older	23.1	23.0	32.2	36.7	42.9	37.2

[a]Internet users at home are individuals aged 16 years old and over who answered they used the Internet from home for personal nonbusiness purposes in the past 12 months. The target population for the Canadian Internet Use Survey has changed from individuals 18 years of age and older in 2005 to individuals 16 years of age and older in 2007.

Source: Statistics Canada, CANSIM, table 358-0128. Last modified: 2009-08-20.

Legislated Privacy

As Woo (2006) correctly pointed out, part of the problem lies in the fact that "legislation is effective only to the degree to which it can be enforced both practically and technologically. The network environment has proven that enforcement is not easy, and formal legislation can prohibit only known or identifiable activities" (p. 960). Thus, we are left with the troubling realization that although we have *de jure* policies in place, the *de facto* reality is that these laws are inadequate, and their application is limited in scope.

The same holds true when we consider the impact that self-regulation by specific industries in the form of codes of practice and/or privacy standards have on safeguarding one's privacy. In fact, as Poullet (2006) noted, "Self-regulation remains a myth" (p. 210). Recently, there has been a concerted effort underway in the United States to enact omnibus legislation dealing with online privacy. In April 2009, Congressman Rick Boucher, head of the House Subcommittee on Communications, Technology and the Internet, announced his desire to see the committee draft an online privacy law before 2010 (Singel, 2009). Similarly, in September 2009, the House Intelligence Technical and Tactical Intelligence Subcommittee sent a letter to President Barack Obama, urging him to appoint a Cybersecurity Coordinator (Oltsik, 2009).

Still, new and existing policies have to "face a technological landscape characterized by rapid and unpredictable evolution" (Poullet, 2006, p. 206). Therefore, it can be argued that in spite of our best efforts (and best intentions), legislation directed at ensuring online privacy is fraught with potential challenges. Perhaps the University of South Australia's (2009) position that "the Internet will be treated as a potentially hostile environment" (Section 2.6) is apt. The most we can hope for is some semblance of control and enforceable consequences that may ensure a degree of personal privacy that online users expect and to which they are entitled. As a result, while policies purport to protect online privacy, they likely fail to accomplish this goal entirely. In a macrolegislative sense, online users are not fully protected; threats to personal privacy are very real and can result in serious consequences.

Besides the inadequacy of current legislation, it is also important to recognize that, as the Canadian Internet Policy and Public Interest Clinic (2007) at the University Ottawa notes, online environments are far from private. In fact,

> Every time you access content or services through the Internet there is the potential for Websites and online services to collect details about your location, your ISP (Internet Service Provider), previous Websites you have visited, your interests, and your computer's technical configuration. (FAQ section)

A 1999 study conducted by Georgetown University is telling in this regard. It reported that in a sampling of 361 websites, 92.8% of the sites collected at least one type of personal identifying information (Stone Analytics, 2003). The key point of this in the context of legal issues is that many asynchronous discussion boards are not anonymous. Names and emails are visible. In fact, by drilling down into the inner mechanisms of the platform, it is often possible to identify the user's Internet Protocol (IP) address. An IP address is a numerical code unique to an individual computer. As well, all Internet websites have a unique IP address in addition to its hosted name (i.e., My_Website.ca).

Mining Student Data

This collection of personal data is perhaps exacerbated in the context of online classrooms. "It is possible that some students are not aware that data about their actions are being recorded" (Anderson & Simpson, 2007, p. 134). Even though students may not be aware of this data mining, those teaching online classes have tremendous power to track student contributions. By simply clicking a mouse, an instructor can have virtually unlimited access to what has transpired in the course. To further complicate matters, most components of an online class are recorded and stored, and one's identity is known to all other members of the group. Thus, anonymity is not easily assured in an online classroom, as posts often have a name attached to them. As Anderson and Simpson (2007) point out, "We are able to track our students' participation, what they read, when they read it and the number of responses they post. We have a permanent record: students' postings are there for us to peruse and to print off" (p. 134). This often covert surveillance can become a potential problem in terms of those postings that may be construed as inappropriately targeting a fellow classmate. Through their online teaching, Anderson and Simpson have encountered situations in which, because of their online comments, students have been harassed through private mail, email, or telephone calls. This situation is indeed troubling and points to the fact that the written word in an online classroom can be captured in perpetuity, with one's identity attached for the official record. As Nissenbaum (1998) pointed out, the tracking of one's activities, regardless of how "carefree" they might be, seems to be almost built into the very fabric of the Internet:

> Information like email addresses, system characteristics, and a trail of network-based activities are not only effortlessly recorded, but easily combined with information from the physical world. In this activity information technology is doubly implicated as it acts as the medium for transactions as well as repository for the information. (pp. 562–563)

As a consequence, "there are no assurances of privacy when we communicate in the virtual classroom" (Palloff & Pratt, 1999, p. 42). Thus, good judgment must be a constant companion of the online student.

In a similar light, students, often in the heat of debate, can share information that might be construed as potentially outside the bounds of acceptable behavior. A case in point surrounds the issue of ethical standards for public school teachers. Section II of the Nova Scotia Teachers Union's *Code of Ethics* states the following:

(a) The teacher should not make defamatory, disparaging, condescending, embarrassing, or offensive comments concerning another teacher.
(b) The teacher shall not make derogatory remarks about the professional competence of another teacher.

These principles are in accordance with the ethical standards in place in most countries. Certainly, teachers who have violated ethical codes have been reprimanded by the courts for their indiscretions, as evidenced, for example, by the 1986 case of *Re Cromer and British Columbia Teachers' Federation*. This litigation, which involved disparaging statements by a teacher (Cromer) against a fellow member of the profession at a public meeting, resulted in the court ruling that Cromer had violated the province's code of ethics.

These types of cases are all too common. In an online classroom, student posts, which are sometimes critical of individual teachers, administrators, schools, or boards, are sometimes submitted with the posting authors and the subjects' names attached. Even in a password-protected classroom, students who criticize other members of the teaching profession could be subject to violations of ethical codes. As well, statements made by students may even be libelous; there have been lawsuits in the United States and the United Kingdom dealing with cyber libel. As Palloff and Pratt (1999) note, "Privacy is virtually impossible to be sure of in electronic communications. The encryption of messages is the most reliable means of privacy, but it is rarely employed in an academic setting" (p. 44). Musings in a face-to-face class, which are often ephemeral, lose this reality in an online world. "Information that is drawn from the physical world is harbored in electronic databases, which give these records the permanence, malleability and transportability that has become the trademark of information technology" (Nissenbaum, 1998, pp. 561–562). Consequently, nothing is lost, and our written contributions may have a lasting impact both in a public as well as private sense. Thus, online communication is not private, and what one posts can have serious repercussions—both ethical and legal ones.

Professors, in their research endeavors, are required by ethics protocols to protect participant data and to destroy that data after a specific number of years. Hard copies of data are to be stored under lock and key. The question remains whether such rigorous protocols are implemented when professors and administrators process and store student data. In an investigation into aligning security policies with procedures, the IT department of one small college in the United States reported that faculty understood rules protecting the confidentiality of student records and the importance of security in protecting their own data (Kuhn, Brookes, & Bellos, 2008). However, as Kuhn et al. noted, stringent security measures may be implemented, but if faculty or students "share passwords, leave printouts in the open, or download copies of data in text files, the privacy of data may be compromised" (p. 74). In an era in which mobile learning and working extends not only to students but also to professors and administrators, the potential to have sensitive files that are stored on portable devices to be misplaced or stolen is a pertinent issue.

CONCLUSIONS

> The online context in which we work, and the interplay of instructor and student rights, engender a range of ethical issues with which we must all engage.
>
> —Anderson & Simpson, 2007, p. 136

So what are we left with? Ultimately, there are laws designed to protect online privacy. However, in their present form, most legislation is incomplete and incapable of delivering absolute privacy to online users. At the same time, a password-protected site is perhaps not as private as we would like to believe. Instructors are equipped with a plethora of tools designed specifically to allow them to keep a watchful eye on their students. The conundrum associated with tracking students online is captured by Anderson and Simpson (2007), who pose the questions:

> [When] does our interest in knowing more about a student—in order to make meaningful links to learning—intrude on the student's right to privacy? Do we need to ask for permission to watch students' online interactions, to review and reassess their online contributions and to "eavesdrop" on their "conversations?" (p. 135)

These queries are difficult to answer, but they bring to light the fact that online class content may not be completely private and can, in fact, be used in suspect ways. Although "computer technology and digital media have . . . increased the capacity to collect, process and use personal information, [they] have also deeply challenged the dynamics surrounding personal information and privacy" (Woo, 2006, p. 953).

As well, individual onus for ensuring the appropriate safety and security of personal data may give rise to differing interpretations of privacy and security laws. For example, the U.S. Family Education Rights and Privacy Act (20 U.S.C. § 1232g; 34 CFR Part 99) contains a list of exclusions and caveats in the nonconsensual release of private student documents to outside agencies; one such exclusion is to "organizations conducting certain studies for or on behalf of the school" (U.S. Department of Education, n.d.). As well, FERPA also outlined policy regarding the publication of student directories and availability of email addresses (U.S. Department of Education, 2008). Yet this document's vague wording opens the door to subjective interpretations and potential misuses of personal data.

RECOMMENDATIONS

Ironically, many companies and learning institutions bury their privacy statements deeply in corporate documents and websites, thus adding further confusion to the entire notion of privacy and transparency. Too often these same privacy statements lack specificity in terms of global applications. To provide much-needed clarity, the implementation of standards in the management of personal data is necessary.

Specific technological policies such as those at Simmon's College (Kuhn et al., 2008) are useful to consider. Managers and course administrators would be well advised to consider the following series of questions:

- Does the application have a security scheme that enforces roles that permit users to do or view only what they are supposed to?
- Does the system require its own set of user names and passwords, or does it integrate with campus authentication schemes?
- What liabilities does the system create (or mitigate), for example with respect to unauthorized access to data?
- Who will maintain this security scheme?
- When personnel change or change roles, how will the security scheme be updated? (Kuhn et al., 2008, p. 73)

Yet having asked questions such as these is not sufficient; clear, definitive answers are also necessary, but perhaps impossible to provide in global contexts.

When implementing appropriate and manageable privacy and security measures in face-to-face and online settings, policies need to extend across technical, financial, and human domains. Privacy policies need to be explicit. The manner in which Kuhn et al. (2008) have outlined the Simmons College Technology department's stance on privacy and security is deceptively simple. They explicated how "the protocol *cautions* people to review system security to ensure that their department abides by institutional standards and legal requirements" (p. 72, emphasis added). By putting the onus on individuals, without appropriate regulation and follow-up, breaches and negligent activities may ensue.

From a business perspective, institutions have yet to meet the increasing demands from consumers for online learning. As technological advances afford more opportunities, students in technical communications programs are able to choose from a wealth of learning opportunities. Through the introduction of new technologies, universities and other learning organizations are better able to compete in a global education market and tap into alternate sources of revenue. The question is not whether organizations will offer online learning opportunities but rather whether they will take the time to do it well (Rosenburg, 2001, p. xvi) and to implement effective policies and procedures that guide and protect online content and interaction. The sheer growth of e-learning opportunities in Canada, the United States, and across the globe is perhaps daunting, yet that growth and enthusiasm needs to be tempered with realism and pragmatics.

REFERENCES

Abrami, P. C., Bernard, R. M., Wade, C. A., Schmid, R. F., Borokhovski, E., Tamin, R., et al. (2006). *A review of e-learning in Canada: A rough sketch of the evidence, gaps, and promising directions*. Montreal, Canada: Centre for the Study of Learning and Performance, Concordia University.

Abramowicz, A. (2006). Software solutions and PIPEDA compliance. *Oral Health, 96*(8), 57–61.

Anderson, B., & Simpson, M. (2007). Ethical issues in online education. *Open Learning, 22*(2), 129–138.

Bloom, M., & Murray, D. (2001). *E-learning for the workplace: Creating Canada's lifelong learners.* Ottawa, Canada: Conference Board of Canada. Retrieved August 15, 2011, from http://www.conferenceboard.ca /documents.aspx?DID=62

Boyle, J. (1996). *Shamans, software, and spleens.* Cambridge, MA: Harvard University Press.

Canadian Council on Learning. (2009). *The state of e-learning in Canada.* Ottawa, Canada: Author. Retrieved August 15, 2011, from http://www.ccl-cca.ca/pdfs/E-learning/E-Learning_Report_FINAL-E.PDF

Canadian Internet Policy and Public Interest Clinic. (2007). *Privacy.* Retrieved October 10, 2009, from http://www.cippic.ca/privacy/

Charpentier, M., Lafrance, C., & Paquette, G. (2006). *International e-learning strategies: Key findings relevant to the Canadian context.* Victoria, Canada: Canadian Council on Learning. Retrieved August 15, 2011, from http://www.ccl-cca.ca/pdfs/OtherReports/JohnBissInternationalELearningEN.pdf

Council of Ministers of Education, Canada. (CMEC). (2001). *Vision statement on on-line learning in postsecondary education.* Retrieved from http://www.cmec.ca

Curran, C. (2008). Online learning and the university. In W. B. Bramble & S. K. Panda (Eds.), *Economics of distance and online learning: Theory, practice, and research* (pp. 26–51). New York, NY: Routledge.

DeCew, J. W. (1997). *In pursuit of privacy: Law, ethics & the rise of technology*. Ithaca, NY: Cornell University Press.

Dell, C. A., Hobbs, S. F., & Miller, K. (2008). Effective online teacher preparation: Lessons learned. *MERLOT Journal of Online Learning and Teaching, 4*(4), 602–610.

European Centre for the Development of Vocational Training (ECDVT). (2001). *E-learning and training in Europe: A survey into the use of e-learning in training and professional development in the European Union.* Luxembourg: Office for Official Publications of the European Communities. Retrieved from http://www.refernet.org.cy/Publications/Cedefop/Pdfs/Surveys/Ced3021.E-Learn&Train.Europe.02.EN.pdf

European Digital Rights. (2008, April 23). Eurobarometers on data protection in EU. *EDRI-gram, 6*(8). Retrieved from http://www.edri.org/edrigram/number6.8/eurobarometer-data-protection

export-gov. (n.d.). *European Union Safe Harbor overview*. Retrieved from http://www.export.gov/safeharbor/eu/eg_main_018476.asp

Fried, C. (1968). Privacy. *Yale Law Journal, 77*(3), 475–493.

Gerety, T. (1977). Redefining privacy. *Harvard Civil Rights-Civil Liberties Law Review, 12*(2), 233–293.

Grant-Davie, K., & Cargile Cook, K. (2005). Introduction. In K. Cargile Cook & K. Grant-Davie (Eds.), *Online education: Global questions, local answers* (pp. 1–14). Amityville, NY: Baywood.

Information Commissioner's Office. (n.d.). *The basics*. Retrieved from http://www.ico.gov.uk/what_we_cover/data_protection/the_basics.aspx

Instructional Technology Council. (2008). *2008 distance education survey results: Tracking the impact of elearning at community colleges*. Retrieved from http://www.itcnetwork.org /file.php?file=%2F1%2FITCAnnualSurveyMarch2009Final.pdf

Kuhn, R. M., Brookes, K. H., & Bellos, N. (2008). A policy-based approach to technology systems. *EDUCAUSE Quarterly, 31*(2), 70–74.

McCullagh, K. (2008). Blogging: Self presentation and privacy. *Information and Communications Technology Law, 17*(1), 3–23.

McKeown, L., & Underhill, C. (2005). Learning online: Factors associated with use of the Internet for educational purposes. Ottawa, Canada: Statistics Canada. Retrieved from http://www.statcan.gc.ca/pub/81-004-x/2007004/10375-eng.htm

Minister of Justice. (2009). *Personal information protection and electronic documents act*. Ottawa, Canada: Author. Retrieved from http://laws.justice.gc.ca/PDF/Statute/P/P-8.6.pdf

Newswire Today. (2009, August 25). 2009 EducationDynamics eLearning Index reveals new supply and demand gaps in online education market. *Newswire Today*. Retrieved from http://www.newswiretoday.com/news/56105/

Nissenbaum, H. (1998). Protecting privacy in an information age: The problem of privacy in public. *Law and Philosophy, 17*(5/6), 559–596.

O'Leonard, K. (2009). *Corporate learning factbook 2009: Executive summary*. Bersin & Associates. Retrieved from http://www.bersin.com/uploadedFiles /010909_ES_Factbook2009_KOL_Final.pdf

Oltsik, J. (2009, October 6). Obama pressed (again) on cybersecurity. *PCWorld*. Retrieved from http://www.pcworld.com/article/173223 obama_pressed_again_on_cyber security.html

Organization of Economic Co-Operation and Development (OECD). (n.d.). *OECD guidelines on the protection of privacy and transborder flows of personal data*. Retrieved from http://www.oecd.org/document/18/0,2340,en_2649_34255_1815186_1_1_1_1,00.html

Palloff, R. M., & Pratt, K. (1999). *Building learning communities in cyberspace: Effective strategies for the online classroom*. San Francisco, CA: Jossey-Bass.

Poullet, Y. (2006). The directive 95/46/ED: Ten years after. *Computer Law & Security Report, 22*, 206–217.

Rosenberg, M. J. (2001). *E-learning: Strategies for delivering knowledge in the digital age*. New York, NY: McGraw-Hill.

Ruth, S. R., Sammons, M., & Poulin, L. (2007). E-learning at a crossroads—What price quality? *EDUCAUSE Quarterly, 30*(2).

Singel, R. (2009, April 23). Online privacy law on the way, congressman promises. *Wired Online*. Retrieved from http://www.wired.com/epicenter/2009/04/netonline-priva

Sloan Consortium (Sloan-C). (2008). *Home page*. Retrieved from http://www.sloan-c.org/

St.Amant, K. (2007). Online education in an age of globalization: Foundational perspectives and practices for technical communication instructors and trainers. *Technical Communication Quarterly, 16*, 13–30.

Statistics Canada. (2008). *Canadian Internet use survey, 2007: Questionnaire*. Retrieved from http://www.statcan.gc.ca/imdb-bmdi/instrument/4432_Q1_V7-eng.pdf

Statistics Canada. (2009). *Internet use by individuals, by type of activity*. Retrieved from http://www40.statcan.gc.ca/l01/cst01/comm29a-eng.htm

Stone Analytics. (2003). *Georgetown Internet Privacy Policy Survey: Executive summary*. Retrieved from http://www.secondmoment.org/etal-column/georgetown.php

University of South Australia. (2009). *Security and information policy*. Retrieved from http://www.unisa.edu.au/ists/GovernanceInIT/Policies/others/security.asp#Internet

U.S. Department of Education. (n.d.). *Family Educational Rights and Privacy Act (FERPA): General*. Retrieved from http://www.ed.gov/policy/gen/guid/fpco/ferpa/index.html

U.S. Department of Education. (2008). *Part II: Department of Education 34 CFR Part 99. Family educational rights and privacy; Final rule*. Retrieved from http://www.ed.gov/legislation/FedRegister/finrule/2008-4/120908a.pdf

Warren, S. D., & Brandeis, L. D. (1890). The right to privacy. *Harvard Law Review, 4*(5), 193–220.

Weigel, V. B. (2002). Commoditization of instruction. In *Deep learning for a digital age: Technology's untapped potential to enrich higher education* (pp. 30–59). San Francisco, CA: Jossey-Bass.

Woo, J. (2006). The right not to be identified: Privacy and anonymity in the interactive media environment. *New Media & Society, 8*(6), 949–967.

http://dx.doi.org/10.2190/LIIC2

CHAPTER 2

The Power of Slogans: The Rhetoric of Network Neutrality

Brett Lunceford

The technological landscape that we inhabit is becoming increasingly complex, yet citizens and government officials are often uninformed concerning the intricacies of these technologies. Ellul (1992) asks, "How can people who are incompetent make important decisions with regard to technique? Here, of course, ordinary citizens are in exactly the same place as the politicians, who are also perfectly incompetent" (p. 43). Without an understanding of how technology works, citizens and government officials are ill-equipped to make rational, well-informed decisions concerning its implementation and governance. This situation can lead to poor policy at best and complete miscarriages of justice at worst. Because of the difficulty in explaining complex technological systems to laypersons, technologies are often reduced to simplistic explanations and slogans. The problem here is that in the case of network neutrality, technical communicators have largely chosen to reinforce such oversimplification by jumping onto one bandwagon or another instead of working to clarify the issues.

This chapter examines the current debate surrounding network neutrality—a situation in which the technical details have been largely ignored and slogans and arguments that appeal to heuristics have come to dominate public perceptions. Despite the complexity of the issues surrounding Internet traffic in international contexts, proponents of network neutrality have centered the debate on the master term of neutrality. Such an argument implies that the issue here is fairness, and herein is the problem. The argument is no longer a technical question; it has become a moral issue bound up with emotion, and Cicero's (1942 version) observation that "men decide far more problems by hate, or love, or lust, or rage, or sorrow, or joy, or hope, or fear, or illusion, or some other inward emotion, than by reality, or authority, or any legal standard, or judicial precedent, or statute" is just as valid in the modern public sphere as it was in the Roman assembly (p. 325). The technical communicator must serve as a logically grounded counterpoint to the rhetoric of doom and fear that is often prevalent in the public sphere.

For technical communicators, the situation has become one in which it is easy to be led away by flowery speech and seductive metaphors. As a result, technical communicators need to understand how language shapes our perceptions of technologies. After all, network neutrality is a complex issue that encompasses packet management, bandwidth allocation, and the right to manage one's network, rather than simply a black-and-white case of freedom versus oppression. Moreover, one must recognize that the term *technical communicator* encompasses not only traditional roles such as technical writers and editors, but also those who create, market, and distribute technologies and the policies that surround them. In short, anyone who interprets technologies to the public is a technical communicator; including technology scholars, journalists, intellectual property attorneys, inventors, and anyone else who has a vested interest in technology adoption and policy. For the purposes of this chapter, I will proceed with this broad conception of the technical communicator.

The case of network neutrality provides a case study in which to consider how technology issues are rhetorically constructed in the public sphere. These constructions can have severe implications for law and public policy. That is, when public policy is created out of emotion rather than a clear understanding of the technology in question, it is the citizenry as a whole that will lose. As such, the technical communicator plays a vital role in helping society understand technologies and their impact. Moreover, such legislation may also have global consequences, especially in the case of communication technologies that span the globe. For example, in an age of widely available, easy to use digital reproduction, copyright law would mean little without international copyright treaties. This essay examines these issues and then concludes with suggestions for how technical communication professionals can help to clarify debates surrounding technologies and their implications for laypersons, legislators, and legal practitioners.

THE PROBLEM WITH DISTRIBUTED NETWORKS

The Internet, as it exists in the public mind, is a fiction. Even the words we use to describe the Internet portray it as a monolithic entity—*The* Internet. But the Internet is not so much an entity as it is a system of smaller networks that their administrators have chosen to link together, much as the highway system is simply a system of interconnected roads. As a network of networks, the Internet is the epitome of decentralization. This was, of course, by design; when ARPAnet, the precursor to the Internet, was built by ARPA (Advanced Research Project Agency), it was meant to withstand a nuclear attack on the United States (Salus, 1995; Thomas, 2002).

Yet as with many technological innovations, the Internet comes with many unintentional consequences, not least of which is the difficulty in controlling a network that, by design, resists control. The Internet has given rise to vexing legal and ethical issues, including

- Identity theft (Allison, Schuck, & Lersch, 2005; Eisenstein, 2008; Marshall & Tompsett, 2005; Simpson, 2003)
- Online gambling (Morse, 2007; Whybrow & Reed, 2002)
- Pharmaceutical sales (Binns & Driscoll, 2001; Fox, Ward, & O'Rourke, 2005; Levaggi et al., 2009)
- Hyperlinking (Auld, 2001)
- Pop-up ads (Nosko, Wood, & Desmarais, 2007; Tyacke & Higgins, 2005)
- Auction fraud (Chua & Wareham, 2008; Jenamani, Zhong, & Bhargava, 2007; "US fraud victims," 2007)
- Child pornography and exploitation (Bauserman, 2003; Beech, Elliott, Birgden, & Findlater, 2008; Burke, Sowerbutts, Blundell, & Sherry, 2002; Schell, Martin, Hung, & Rueda, 2007; Williams, 2003)
- Intellectual property piracy (Chiang & Assane, 2008; B. Lunceford & Lunceford, 2008; Peitz & Waelbroeck, 2006; Seay, 2008; Sims, 2003; Yang, 2007)
- Various forms of email fraud (Chua & Wareham, 2008; Edelson, 2003; Ojedokun, 2005; Oriola, 2005).

The Internet allows each and all of these actions to take place on a global scale, and as more nations gain access to the Internet, the scope and scale of these activities will only continue to grow and to globalize. Moreover, these issues lay bare sometimes opposing attitudes held by different nations concerning such issues as

- Privacy (Baumer, Earp, & Poindexter, 2004; Beldad, De Jong, & Steehouder, 2009; Brown & Blevins, 2002; Cheung, 2009; Floridi, 2005; Poindexter, Earp, & Baumer, 2006; Wang & Hong, 2010)
- Freedom of speech (Abbott, 2001; Chin, 1997; Dickerson, 1996; Leets, 2001; Sunstein, 1995; Yang, 2007)
- Sexual morality (Baurmann, 2004; Graupner, 2004a, 2004b; Hofmeister, 2004; Waites, 2005).

Once again, the global nature of the Internet allows for these clashes in values because what may be perfectly acceptable in one country might be against the law in another. As more and more nations become increasingly wired, these clashes are likely to become more frequent. Technical communicators will play an important role not only in identifying these potential clashes but in mediating the controversies that will inevitably arise.

In essence, these factors reveal a simple but important truth. Because of its decentralized nature, the Internet can no longer be controlled only by those who created it. Rather, the genie is out of the bottle, and the Internet now connects the entire globe. As such, any effort to regulate the Internet by one entity is likely to be thwarted by the efforts of another. For technical communicators, these developments mean that one can no longer resort to simplified discussions of technologies and their impacts. Despite the apparent difficulty of explaining technologies to laypeople, the increasing complexities of technological systems demand complex explanations that accentuate, rather than deny, the nuances of these systems.

The Nature of the Internet

When considering the implications for Internet regulation, we must first begin with the nature of the Internet itself, the code itself. Lessig (1996) notes that "Engineers write the code; the code defines the architecture, and the architectures define what is possible within a certain social space" (p. 1410). According to Lessig,

> Code is an efficient means of regulation. But its perfection makes it something different. One obeys these laws as code not because one should; one obeys these laws as code because one can do nothing else. . . . Law as code is a start to the perfect technology of justice. (p. 1408)

Yet code is not simply a mechanism for control. Those who write the code endow it with certain attributes, based on the values of the creator. As such, the code is also a means of distributing an ideology. Hacktivists especially have noted that the architecture of the Internet allows for a level of direct action even in oppressive governments that would be difficult to attain in the non-virtual world (Jordan, 2002; Jordan & Taylor, 2004).

What is at stake in the debates concerning network neutrality is the architecture of the Internet itself. Ganley and Allgrove (2006) note that the Internet "was designed as a 'dumb' network" (p. 456). In other words, everything that passed over the network is treated just like anything else—a packet is a packet is a packet, regardless of what that packet contains. However, Ganley and Allgrove explain that routers allow operators "to prioritize or de-prioritize certain packets of data or even drop them from their network altogether" (p. 454). The question, then, is whether network administrators should be allowed to prioritize some packets over others. Network neutrality advocates (e.g., Herman, 2006; Lessig, 2007) argue that the "end-to-end" design of the dumb pipe is essential to maintaining innovation online, while others, such as Hass (2007) argue that the

Internet never was neutral, that even from the beginning the ability to prioritize packets was built into the network.

It seems that if technical communicators could provide a technical description of the Internet as it is and has been, the public would be able to more fully recognize the inherent flaws in the arguments of those who have chosen to engage in a kind of historical revisionism to advocate for a "dumb pipe" that never was. However, this would require technical communicators to eschew the metaphor-driven language of the network neutrality debate and instead adopt the language of science. Such a move, however, would require technical communicators to consider ways to make scientific language more accessible to the layperson.

Network Neutrality and the Law

Legal scholars have discussed the legal implications of network neutrality, including specific pieces of legislation (Beard, Ford, Koutsky, & Spiwak, 2006; Burstein & Schneider, 2009; Frieden, 2007; Herman, 2006; Lessig, 2007; S. Lunceford, 2008; Wong & Garrie, 2007; Yoo, 2005a), while others have looked at it from an economic perspective (Fischmann & van Schewick, 2006; Hermalin & Katz, 2007; Shrimali, 2008; Yoo, 2005b). S. Lunceford (2008) argues forcefully that proposals for network neutrality are unsustainable from a networking standpoint. As a legal scholar and a computer scientist, Lunceford lays out both the legal arguments against network neutrality as well as the technical ones based on the architecture of the Internet and networking protocols. Yet, as I have explained elsewhere, the construction of law is, at its heart, a rhetorical concern (B. Lunceford & Lunceford, 2008), so for the remainder of this chapter, I will examine the rhetorical contours of the network neutrality debate in order to consider how this issue is discussed in the public sphere. In this case, we will see that by allowing the network neutrality camp to define the terms that describe the technology, many technical communicators have abdicated their responsibility to help others understand technological systems. Moreover, by allowing a metaphor- and ideology-driven language to define these systems, they have unwittingly helped to define network neutrality as an unquestioned good, despite serious technical flaws. As such, the case of network neutrality serves as a cautionary tale of how language shapes our perceptions of technologies and technological systems.

THE RHETORIC OF NETWORK NEUTRALITY

Rhetorician Thomas Benson (1985) explains, "Rhetoric critics inquire into meaning, not simply in an artifact but also in the pragmatics of that artifact: that is, in how a human being can, or did, or should use that artifact" (p. 204). Lucas (1988) echoes this sentiment:

> The benefit of close textual analysis is that it allows the critic, in essence, to "slow down" the action within the text so as to keep its evolving internal context in sharp focus and to allow more precise explication of its rhetorical artistry. (p. 249)

To that end, I wish to focus on specific terms in the debates surrounding network neutrality, carefully considering how these terms function to invite the observer to understand the issue.

First, we must consider who the players are in the debate. Ganley and Allgrove (2006) observe that "the net neutrality debate is often framed as having just two sides. On one side are the operators. . . . The other side of the debate is more complex and is characterized by an eclectic coalition of content and service providers" (p. 455). Although I agree that the debates surrounding network neutrality have many more participants, for our purposes, we can maintain the illusion of two camps while acknowledging the loss of nuance in the arguments.

Each side maintains a belief in specific key ideals. McGee (1975) writes, "Each political myth presupposes a 'people' who can legislate reality with their collective belief. So long as 'the people' believe basic myths, there is unity and collective identity" (p. 245). The ancient Greeks called this common knowledge *doxa*, and such common sense is often bound up in what Weaver (1953) calls "God" and "Devil" terms. These ideals guide the thoughts and actions of collectives—nations, religions, or social movements—that share a particular ideology. As McGee (1980) observes, "Human beings are 'conditioned,' not directly to belief and behavior, but to a vocabulary of concepts that function as guides, warrants, reasons, or excuses for behavior and belief" (p. 6). In other words, the terms that we use to define objects or phenomena shape the audience's perceptions of them. As such, technical communicators must carefully consider the terms that they use to describe technologies.

Master Terms in the Network Neutrality Debate

The network neutrality camp has skillfully seized the reins of the public argument surrounding network neutrality by defining their cause in such terms as *freedom*, *openness*, and *saving the Internet.* I suggest that these terms function as *ideographs*, or terms that are laden with ideology (McGee, 1980). McGee (1980) notes that one cannot argue with the fundamental logic of an ideograph (p. 7), and terms such as *freedom* and *openness* are specifically Western values. Yet when one considers the nature of the Internet, how is it that the United States can actually "save the Internet?" It seems that because the Internet is an international network, any effort to save it would have to be an international effort. Network neutrality advocates seem to be continuing the trend of the United States to go it alone rather than work toward an international solution to the perceived problem.

There are precedents for such international agreements. For example, the World Intellectual Property Organization (WIPO) is an international organization that works to protect intellectual property. Such an organization is necessary (in the minds of copyright advocates at least) because other countries may not value intellectual property in the same way, making it easy to circumvent intellectual property protections by simply doing so in a country that allows one to do so. It came as little surprise that Australia-based Sharman Networks, the creators of the popular Kazaa peer-to-peer file sharing program, incorporated in the island nation of Vanuatu; Vanuatu is not a member of WIPO. Technical communicators must recognize the nuances in arguments surrounding technology and how the various parts of technological systems fit together, both within one's own nation and beyond.

Network Neutrality as "Freedom"

Let us now consider the two terms most prominent in the discourse of network neutrality advocates: freedom and neutrality. In the United States, it is difficult to argue with the term *freedom* because it is such a core foundational belief. The term evokes such fundamental ideals as "*freedom* of speech," "life, *liberty*, and the pursuit of happiness," and "*freedom* of choice." Some have noted that the Internet is a site of freedom; Jewkes and Sharp (2002) argue that the Internet "can liberate its users from the usual constraints of corporality. The Internet thus gives users a freedom of expression—a freedom of *being*—quite unlike anything they have at their disposal in the physical world" (p. 3). By tying the cause of network neutrality to the ideal of freedom, proponents have evoked the idea of freedom of information, freedom of movement, and, most importantly, freedom from control. They paint the alternative to neutrality, and thus freedom, in terms of control by faceless corporate conglomerates bent only on maximizing profits.

That anyone could be an enemy of freedom may seem absurd, but the idea has gained traction in contemporary political discourse, especially as it relates to the online world. Tom Ridge, then-Secretary of Homeland Security, was quoted in the St. Louis *Post Dispatch*:

> Terrorists know that a few lines of code could ultimately wreak as much havoc as a physical attack . . . The enemies of freedom use the same techniques as hackers do . . . and we must be as diligent and determined as the hackers. (Associated Press, 2003, p. B2)

In any movement, it is important to define oneself as among those who are on the side of goodness, righteousness, and, in this case, freedom. It is just as important to define one's enemy as those evildoers who wish to deny freedom, righteousness, and justice. By placing these terms in opposition, proponents of network neutrality, then, invite the listener to make the connection between freedom and slavery or oppression. This can be a powerful strategy both as directed at the public in general as well as in affirming one's own conviction of righteousness. Gregg (1971) notes, "By painting the enemy in dark hued imagery of vice, corruption, evil, and weakness, one may more easily convince himself of his own superior virtue and thereby gain a symbolic victory of ego-enhancement" (p. 82). Or, as Eric Hoffer succinctly argued, "Mass movements can rise and spread without belief in a God, but never without belief in a devil" (1951, p. 91).

Networks as "Neutral"

Neutrality seems somewhat at odds with American values. However, when viewed in the context of network neutrality, the term becomes synonymous with *fairness* and *equality*. Thus, the neutrality evoked is not the neutrality of a nation-state, but rather the impartiality of a judge, an imperative to be no respecter of persons. This is, in theory, an admirable proposal. However, when prodded a bit, it becomes clear that neutrality as a key term is unsustainable because the truth of the matter is that not all traffic *should* be considered equal.

To illustrate this point, consider the following scenario. In many metropolitan areas, there are carpool lanes built into the freeway system because we place a higher value on those who commute together in an effort to reduce congestion within the system. Now, let us imagine that the carpool lane is gone and that there is only one lane. Should four people commuting to work together in a sensible vehicle be seen as equal to a lone 16-year-old driving a Hummer loaded with nothing but cheap beer and pornography? We consider that the car full of people commuting to work provides more value to society as a whole and thus should be given priority. Likewise, the idea that all Internet traffic should be considered equal is ludicrous. It is difficult to assert that BitTorrent traffic, which is often used to download large quantities of copyrighted material such as movies, should be considered at the same level of importance as voice over IP traffic. Yet such is the power of a convincing narrative. If one considers the idea of neutrality as a desirable good, then he or she is less likely to consider fully the implications of the ideographs used to describe the phenomenon. The slogan becomes the whole of the argument.

Technical communicators must recognize that it is not always possible to map a particular social value from the physical world onto the digital one, especially when one ignores the implications of those values. For example, as a society, the United States places a high value on freedom of speech—so much so that it is enshrined in the First Amendment. Yet there are some assumptions that are often overlooked in the First Amendment that prove problematic in cyberspace. For example, in physical space, the speech is often connected to a person, and the person generally will claim such speech (unless he or she is a coward). In cyberspace, however, it is quite easy to anonymously post libelous comments. This calls attention to the nature of being (in the ontological sense) in the online world; Thomas (2002) explains that in order to try someone for a crime, a virtual presence is not enough and that "what is always needed is a body, a real body, a live body" (p. 182). Moreover, if one takes the idea of freedom of speech at face value, it is easy to ignore the fact that these posts are often not posted in public space, wherein freedom of speech is most powerfully upheld, but rather on the private property of another's server space. As such, cries of censorship that often arise when a moderator deletes an offensive posting is actually much more akin to a homeowner removing graffiti from his or her home than a government entity quashing free expression. When a technical communicator advocates for one side or another, he or she must understand the full scope of how the technology is used.

INTERNATIONAL IMPLICATIONS FOR NETWORK NEUTRALITY LEGISLATION

The Internet is only as good as the connections that make it up. By seeking network neutrality legislation, the U.S. government seems intent on establishing worldwide Internet law. Yet in some cases, this is already occurring. Braman and Roberts (2003) argue that Internet Service Providers' (ISP) Acceptable Use Policy and Terms of Service agreements are becoming a kind of de-facto law because governments worldwide are demanding more of ISPs, placing them into a regulatory role, yet allowing ISP regulations that may not be constitutional. Moreover, Nunziato (2003) provides a detailed explanation of overlooked unintended consequences of information policy in her discussion of the Internet Corporation for Assigned Names and Numbers (ICANN), arguing that ICANN policies restrict freedom of speech, specifically anonymous and/or critical speech. Nunziato also notes that ICANN's policies make it difficult to appeal to the U.S. court system because there is only a window of 10 days to lodge a court appeal before the decision is considered final and binding (p. 212).

American political wranglings concerning network neutrality often seem to overlook the international implications of such legislation. Such an attitude is understandable in light of Schiller and Miège's (1990) assertion that

> The "information society," as it now functions, takes for granted, indeed reinforces, a world economy with an international division of labor that apportions benefits to a small number of highly developed nations. These are then redistributed, also unequally, within these privileged economies. (p. 163)

Yet such a mentality must be altered in light of the increasingly wired world, even in developing nations (Borzekowski, Fobil, & Asante, 2006; Ford, 2007;

Hong & Huang, 2005; Lau, Aboulhoson, Lin, & Atkin, 2008; Mwesige, 2004; Robison & Crenshaw, 2002; Wang & Hong, 2010).

Ganley and Allgrove (2006) note that "the net neutrality debate is one which is now gaining traction in Europe" (p. 455). However, de Bijl and Peitz (2008) observe that "while there has been a heated policy debate on network neutrality in the US, it seems that Europe is lagging behind" (p. 750). Differences in regulatory structures from those found in the United States have played a part in debates over network neutrality in Europe. For example, Cave and Crocioni (2007) state that "the degree of access regulation for Internet broadband in the US is currently considerably lower than in Europe where often because of access obligations, the retail ISP is not the wholesale network provider" (p. 670). In fact, because of existing regulatory frameworks in Europe, Cave and Crocioni conclude that "the need for a 'net neutrality' provision is, at best, not proven in Europe" (p. 678). Even so, because of the interconnected nature of the Internet, the regulations forged in the United States are likely to have an impact on the rest of the world. Yet outside of the academic press, such cross-cultural implications are often ignored in the debates surrounding network neutrality.

In an international context, network neutrality is troubling from an economic point of view, especially in countries that do not have a desirable consumer base. Some content companies have decided to limit or even eliminate some markets because of financial concerns. For example, the San Diego-based company Veoh, a website for video sharing, cited a lack of prospects for profit generation, in combination with the relatively high cost of delivering video, as primary reasons behind its decision to block the access users in Africa, Asia, Latin America and Eastern Europe had to its services. As one source noted,

> "I believe in free, open communications," Dmitry Shapiro, the company's chief executive, said. "But these people are so hungry for this content. They sit and they watch and watch and watch. The problem is they are eating up bandwidth, and it's very difficult to derive revenue from it." (Stone & Helft, 2009)

For better or for worse, Internet access is still a business concern. Although it is noble to believe in universal access, there is always the question of who is willing to pay for such access. Moreover, there are other competing interests, such as national security; Winseck (2008) observes that "open networks have been traded off for national security and greater commercial control" (p. 433).

Technical communicators can play a vital role in helping the public understand the nuances of technological systems, but they seem to have largely dropped the ball on the issue of network neutrality. The debate has been framed largely as a battle between the forces of equality and oppressive corporate interests that want to clamp down on the Internet. Rather than challenge the terminology put forth by proponents of network neutrality, technical communicators have generally adopted the terms of neutrality. This painted opponents of network neutrality into a rhetorical corner, and it seems that technical communicators would have done well to eliminate metaphorical language and instead focus on the technical aspects of the system.

IMPLICATIONS FOR TECHNICAL COMMUNICATORS

As the case of network neutrality demonstrates, the potential for ambiguity is present even when parties share a culture and a language. When the stakeholders in a technology such as the Internet span the globe, it becomes even more important to communicate in ways that are as accurate as possible. To that end, I have outlined four strategies for technical communicators: keep it technical; choose words carefully; consider the big picture; and consider the international audience.

Strategy 1: Keep it Technical

Every technical communicator has repeatedly heard that the key is to "Keep it simple." I suggest that such a heuristic has done a disservice to the technical community. In our quest for simplicity, we have overlooked the idea of *elegant* simplicity. In other words, "simple" need not mean "dumbed down." The case of network neutrality demonstrates the dangers of using a too simple explanation for a very complex system. I propose that technical communicators should focus on keeping technical communication technical. By this, I do not mean that technical communicators should make technical issues inaccessible to the layperson. Rather, technical communicators should focus on the technology itself, divorced from the often colorful analogies used to describe them.

Perhaps technical communicators hold to the idea that technical systems are too complicated to be adequately comprehended by the layperson. As Winner (1977) notes, "One becomes accustomed to the idea that systems are too large, too complex, and too distant to permit all but experts an inside view" (p. 288). However, if technical communicators succumb to the temptation to oversimplify complex systems, they are likely to describe these systems in ways that are inaccurate. This can be particularly damaging to one's credibility, even if the desire was to interpret the system in an accessible manner.

When using technical language, technical communicators can more easily avoid language that evokes unintended consequences. For example, Lakoff and Johnson (1980) describe how many metaphors place us in a particular mindset (e.g., metaphors that elicit imagery of war). Other times, technical communicators resort to using stories as examples. However, these narratives often function in ways that were unintended by the author, especially if the reader or listener takes certain aspects of the narrative too literally. Fisher

(1984, 1987) suggests that the listener processes narratives not according to the traditional rules of logic but rather based on whether the story conforms to one's sense of narrative fidelity. In other words, the technical elements can be overshadowed by the story itself.

Strategy 2: Choose Words Carefully

The words one chooses to describe a situation can have drastic consequences. Whorf (1956) reports on many cases of fire and explosions that he was called to investigate, suggesting that the root cause was not merely accidental but also how the individuals involved constructed the situation linguistically. For example, when handling drums filled with gasoline, workers would handle them with care, but with the empty drums, they were "careless, with little repression of smoking or of tossing cigarette stubs about. Yet the 'empty' drums are perhaps the more dangerous, since they contain explosive vapor" (p. 135). Of course, the term *empty* is the real problem in this example, but Whorf points out that although they are not filled with gasoline, they are in no way empty. Whorf goes on to recount several instances in which the words individuals used to describe their situation led to dangerous situations.

Burke (1966) also suggests that the words that we use filter our perception, calling this idea "terministic screens": "Pick some particular nomenclature, some one terministic screen. . . . That you may proceed to track down the kinds of observations implicit in the terminology you have chosen, whether your choice of terms was deliberate or spontaneous" (p. 47). In the case of network neutrality, there are many alternate terms that could be used. For example, one could refer to it as "bandwidth *management*," "packet *discrimination*," or "Internet *freedom*." It is easy to see how these terms color perceptions of the topic at hand.

In addition to the ways in which terms color our perceptions, there is also the danger in abstracting such complex issues as network neutrality in non-technical ways. For example, when using the term *network neutrality*, different people have different conceptions of what the term means, yet they all think that they support the idea. Aristotle (1991 translation) alluded to such an idea in his discussion of the enthymeme, in which the hearer will supply the missing premises from commonly held opinions (pp. 186–187). Bitzer (1959) argues that because enthymemes are "formed out of premises supplied by the audience, they have the virtue of being self persuasive. Owing to the skill of the speaker, *the audience itself helps construct the proofs by which it is persuaded*" (p. 408). The ideograph of *neutrality* paints the listener into a rhetorical corner. If one seeks neutrality rather than discrimination, they will supply the rest of the argument. Describing technological systems in technical terms can often help technical communicators to avoid such linguistic quandaries.

Strategy 3: Consider the Big Picture

Technological systems are increasingly complex and now often span the globe. What one nation does is likely to affect other nations as well. Technical communicators can no longer afford to adopt a myopic worldview that focuses only on one specific nation's interests. Of course many of these effects are difficult to predict and at times may be unintended. For this reason it is imperative that technical communicators try to remain as objective as possible about the technologies they explain.

As technology evolves, it influences society. Some scholars, such as Ellul (1964), argue that "technique elicits and conditions social, political, and economic change. It is the prime mover of all the rest, in spite of any appearance to the contrary" (p. 133). However, it is easy to forget that how we discuss technologies can sometimes be as important as the technologies themselves. In other words, technologies can be rhetorical. Black (1970) argues that rhetorical discourses imply an ideal auditor, for whom the discourse is designed and that this implied auditor can often be linked to a particular ideology (p. 112). Technical communicators must be careful to avoid creating unintended ideologies in their discourses surrounding technologies.

If technology shapes society, one must likewise consider how technical communicators describe the impact of these technologies on society. Technical communicators play a vital role in involving the public in discussions surrounding technologies. Yet Elliot (1986) suggests that "what we are seeing and what we face is a continuation of the shift away from involving people in society as political citizens of nation states towards involving them as consumption units in a corporate world" (p. 106). By involving the public in issues of public concern, technical communicators are an essential component of democratic society, and Dewey (1991) suggests that a well-informed public is vital to a well-functioning democracy.

Strategy 4: Consider the International Audience

In addition to unintended ideological suggestions, there is also the potential for miscommunication as multiple cultures engage with these technologies and the discourses surrounding them. As Thrush (1993) observes, even if the business is conducted in English, "the use of a particular grammar and vocabulary does not necessarily imply a shared value system, a uniform approach to business transactions, or a common pool of knowledge" (p. 273). Likewise, St.Amant (2005) notes that "increased access, however, does not necessarily mean increased acceptance or use of ideas. Rather, differing expectations related to design can affect the cross-cultural transfers of information" (p. 75). Language is only part of the equation. As Lipus (2006) argues, "Writing for multicultural audiences is a complex task, and knowing the culture can be as important as knowing the language" (p. 81). The potential for slippage in communication is evident even in the most routine interpersonal interactions and therefore particularly important to remember when discussing complex technical issues.

One issue of particular importance to technical communicators is the use of ethnocentric language. Something as simple as the word "us" can reveal a particular bias. Technical communicators must consider other worldviews beyond their own, because the rest of the world is now responding. In their discussion of online forums, Starke-Meyerring, Duin, and Palvetzian (2007) note that even though technical communicators might have previously focused on creating documentation for local audiences, in the modern context of globalization,

> [T]hey must now quickly and effectively engage users from multiple and mixed backgrounds in the various global networks in which they come together. The way technical communicators engage such a forum can mean the difference between a showcase of disgruntled, dissatisfied customers and a fan forum of passionate supporters. (p. 142)

As the world becomes increasingly wired, such communication will likely become increasingly interactive.

CONCLUSION

To say that law is often created in the image of particular ideographs is unremarkable. Take, for example, such acts as the U.S. Patriot Act, the Protect Act of 2003, and the No Child Left Behind Act, which simply through their nomenclature short-circuit any possibility of dissent. After all, who would oppose patriotism? The discourse in favor of network neutrality seems to function in a similar way. By claiming the mantle of freedom and equality, proponents are able to connect the idea of network neutrality with deeply held American beliefs. Yet such arguments seem to ignore the international nature of the Internet, and by appealing mainly to universals such as freedom, obscure the nuances of the issue. Moreover, by appealing to a kind of revisionist history, they make a case for an Internet architecture that never was, and perhaps should not be, especially in light of economic concerns surrounding the service of populations that may be unwilling or unable to financially support such an endeavor.

Technical communicators have an important role in helping the public to understand the nuances in the debates over how technology should be used and regulated. If the public remains ignorant of how technology works, they will be ill-suited to make informed opinions and judgments. Moreover, technical communicators have a particularly important role in helping policymakers understand the intricacies of technology. Although it may be tempting to reduce complex issues such as network traffic to simple slogans such as net neutrality, such impulses should be resisted in the interest of robust debate over how technology should be regulated. When slogans are employed, especially when combined with ideographs that quash dissent, deliberation is impoverished and democracy itself is at risk. After all, knowledge is a key component in deliberative democracy; James Madison's (1865) words, written in 1822, are just as applicable today:

> A popular Government, without popular information, or the means of acquiring it, is but a prologue to a farce or tragedy; or perhaps both. Knowledge will forever govern ignorance; and a people who mean to be their own governors, must arm themselves with the power which knowledge gives. (p. 276)

In the realm of technology, then, technical communicators are the guardians of information and serve as the means by which citizens can make informed judgments.

Lessig (1996) argued against the idea of a system of cyberlaw that is separate from law in the physical world because

> [T]he effects of that place [cyberspace] will never be far removed from this [physical space]. And our understanding of what that place will become is just beginning. We, here, in this world, will keep a control on the development there. As well we should. (p. 1403)

Yet law crafted through ideas and slogans is law that will certainly have unintended consequences in both the virtual and the physical realms, especially when it is built in the image of a history that never was. Thus, as we continue to watch the Internet evolve across the world, it seems that a collaborative approach that includes all of the involved parties around the world is the most prudent direction forward. Technical communicators must guide this process by explaining clearly how technology works to those who would regulate it, especially when such technologies transcend national borders. Otherwise, the detrimental effects of such regulations will be felt not only in the United States, but across the globe.

REFERENCES

Abbott, J. P. (2001). Democracy@Internet.asia? The challenges to the emancipatory potential of the net: Lessons from China and Malaysia. *Third World Quarterly, 22*(1), 99–114.

Allison, S. F. H., Schuck, A. M., & Lersch, K. M. (2005). Exploring the crime of identity theft: Prevalence, clearance rates, and victim/offender characteristics. *Journal of Criminal Justice, 33*(1), 19–29.

Aristotle (1991). *On rhetoric: A theory of civic discourse* (G. A. Kennedy, Trans.). New York, NY: Oxford University Press.

Associated Press. (2003, December 4). Ridge seeks tech companies' support. *St. Louis Post-Dispatch,* p. B2.

Auld, P. (2001). Internet law: The legal classification of websites and liability for hypertext links. *Computer Law & Security Report, 17*(4), 254–256.

Baumer, D. L., Earp, J. B., & Poindexter, J. C. (2004). Internet privacy law: A comparison between the United States and the European Union. *Computers & Security, 23*(5), 400–412.

Baurmann, M. C. (2004). Sexuality, adolescence and the criminal law: The perspective of criminology. *Journal of Psychology & Human Sexuality, 16*(2/3), 71–87.

Bauserman, R. (2003). Child pornography online: Myth, fact, and social control. *The Journal of Sex Research, 40*(2), 219.

Beard, T. R., Ford, G. S., Koutsky, T. M., & Spiwak, L. J. (2006). Network neutrality and industry structure. *Hastings Communications and Entertainment Law Journal, 29,* 149–169.

Beech, A. R., Elliott, I. A., Birgden, A., & Findlater, D. (2008). The Internet and child sexual offending: A criminological review. *Aggression and Violent Behavior, 13*(3), 216–228.

Beldad, A. D., De Jong, M., & Steehouder, M. F. (2009). When the bureaucrat promises to safeguard your online privacy: Dissecting the contents of privacy statements on Dutch municipal websites. *Government Information Quarterly, 26*(4), 559–566.

Benson, T. W. (1985). The rhetorical structure of Frederick Wiseman's *Primate. Quarterly Journal of Speech, 71*(2), 204–217.

Binns, R., & Driscoll, B. (2001). The Internet, pharmaceuticals and the law. *Drug Discovery Today, 6*(9), 452–453.

Bitzer, L. F. (1959). Aristotle's enthymeme revisited. *Quarterly Journal of Speech, 45,* 399–408.

Black, E. (1970). The second persona. *Quarterly Journal of Speech, 56*(2), 111–119.

Borzekowski, D. L. G., Fobil, J. N., & Asante, K. O. (2006). Online access by adolescents in Accra: Ghanaian teens' use of the Internet for health information. *Developmental Psychology, 42*(3), 450–458.

Braman, S., & Roberts, S. (2003). Advantage ISP: Terms of service as media law. *New Media & Society, 5*(3), 422–448.

Brown, D. H., & Blevins, J. L. (2002). The Safe-Harbor agreement between the United States and Europe: A missed opportunity to balance the interests of e-commerce and privacy online. *Journal of Broadcasting & Electronic Media, 46*(4), 565–585.

Burke, A., Sowerbutts, S., Blundell, B., & Sherry, M. (2002). Child pornography and the Internet: Policing and treatment issues. *Psychiatry, Psychology and Law, 9*(1), 79–84.

Burke, K. (1966). *Language as symbolic action*. Berkeley, CA: University of California Press.

Burstein, A. J., & Schneider, F. B. (2009). Trustworthiness as a limitation on network neutrality. *Federal Communications Law Journal, 61*(3), 591–623.

Cave, M., & Crocioni, P. (2007). Does Europe need network neutrality rules? *International Journal of Communication, 1,* 669–679.

Cheung, A. S. Y. (2009). China Internet going wild: Cyber-hunting versus privacy protection. *Computer Law & Security Review, 25*(3), 275–279.

Chiang, E. P., & Assane, D. (2008). Music piracy among students on the university campus: Do males and females react differently? *Journal of Socio-Economics, 37*(4), 1371–1380.

Chin, A. (1997). Making the World Wide Web safe for democracy: A medium-specific First Amendment analysis. *Hastings Communications & Entertainment Law Journal, 19*(2), 309–338.

Chua, C. E. H., & Wareham, J. (2008). Parasitism and Internet auction fraud: An exploration. *Information and Organization, 18*(4), 303–333.

Cicero, M. T. (1942). *De oratore, books I-II* (E. W. Sutton & H. Rackham, Trans.). Cambridge, MA: Harvard University Press.

de Bijl, P., & Peitz, M. (2008). Innovation, convergence and the role of regulation in the Netherlands and beyond. *Telecommunications Policy, 32*(11), 744–754.

Dewey, J. (1991). *The public and its problems*. Athens, OH: Swallow. (Original work published 1927)

Dickerson, D. L. (1996). <Freedom of expression> and cultural meaning: An analysis of metaphors in selected Supreme Court texts. *Communication Law & Policy, 1*(3), 367–395.

Edelson, E. (2003). The 419 scam: Information warfare on the spam front and a proposal for local filtering. *Computers & Security, 22*(5), 392–401.

Eisenstein, E. M. (2008). Identity theft: An exploratory study with implications for marketers. *Journal of Business Research, 61*(11), 1160–1172.

Elliott, P. (1986). Intellectuals, the "information society" and the disappearance of the public sphere. In R. E. Collins, J. Curran, N. Garnham, P. Scannell, P. Schlesinger, & C. Sparks (Eds.), *Media, culture, and society: A critical reader* (pp. 105–115). London, UK: Sage.

Ellul, J. (1964). *The technological society* (J. Wilkinson, Trans.). New York, NY: Vintage.

Ellul, J. (1992). Technology and democracy. In L. Winner (Ed.), *Democracy in a technological society* (pp. 35–50). Dordrecht, The Netherlands: Kluwer.

Fischmann, B. M., & van Schewick, B. (2006). Network neutrality and the economics of an information superhighway: A reply to professor Yoo. *Jurimetrics, 47,* 383–428.

Fisher, W. R. (1984). Narration as a human communication paradigm: The case of public moral argument. *Communication Monographs, 51*(1), 1–22.

Fisher, W. R. (1987). *Human communication as narration: Toward a philosophy of reason, value, and action*. Columbia, SC: University of South Carolina Press.

Floridi, L. (2005). The ontological interpretation of informational privacy. *Ethics and Information Technology, 7*(4), 185–200.

Ford, D. M. (2007). Technologizing Africa: On the bumpy information highway. *Computers and Composition, 24*(3), 302–316.

Fox, N., Ward, K., & O'Rourke, A. (2005). The birth of the e-clinic. Continuity or transformation in the UK governance of pharmaceutical consumption? *Social Science & Medicine, 61*(7), 1474–1484.

Frieden, R. (2007). Internet packet sniffing and its impact on the network neutrality debate and the balance of power between intellectual property creators and consumers. *Fordham Intellectual Property, Media & Entertainment Law Journal, 18,* 633–675.

Ganley, P., & Allgrove, B. (2006). Net neutrality: A user's guide. *Computer Law & Security Report, 22*(6), 454–463.

Graupner, H. (2004a). The 17-year-old child: An absurdity of the late 20th century. *Journal of Psychology & Human Sexuality, 16*(2/3), 7–24.

Graupner, H. (2004b). Sexual consent: The criminal law in Europe and outside of Europe. *Journal of Psychology & Human Sexuality, 16*(2/3), 111–171.

Gregg, R. B. (1971). The ego-function of the rhetoric of protest. *Philosophy & Rhetoric, 4*(2), 71–91.

Hass, D. A. (2007). The never-was-neutral net and why informed end users can end the net neutrality debates. *Berkeley Technology Law Journal, 22,* 1565–1635.

Hermalin, B. E., & Katz, M. L. (2007). The economics of product-line restrictions with an application to the network neutrality debate. *Information Economics and Policy, 19*(2), 215–248.

Herman, B. D. (2006). Opening bottlenecks: On behalf of mandated network neutrality. *Federal Communications Law Journal, 59*(1), 103–155.

Hoffer, E. (1951). *The true believer: Thoughts on the nature of mass movements.* New York, NY: Perennial.

Hofmeister, L. (2004). 14 to 18 year olds as "children" by law? Reflections on developments in national and European law. *Journal of Psychology & Human Sexuality, 16*(2/3), 63–70.

Hong, J., & Huang, L. (2005). A split and swaying approach to building information society: The case of Internet cafes in China. *Telematics and Informatics, 22*(4), 377–393.

Jenamani, M., Zhong, Y., & Bhargava, B. (2007). Cheating in online auction—Towards explaining the popularity of English auction. *Electronic Commerce Research and Applications, 6*(1), 53–62.

Jewkes, Y., & Sharp, K. (2002). Crime, deviance and the disembodied self: Transcending the dangers of corporeality. In Y. Jewkes (Ed.), *Dot.cons* (pp. 1–14). Portland, OR: Willan.

Jordan, T. (2002). *Activism!: Direct action, hacktivism and the future of society.* London, UK: Reaktion.

Jordan, T., & Taylor, P. A. (2004). *Hacktivism and cyberwars: Rebels with a cause?* New York, NY: Routledge.

Lakoff, G., & Johnson, M. (1980). *Metaphors we live by.* Chicago, IL: University of Chicago Press.

Lau, T. Y., Aboulhoson, M., Lin, C., & Atkin, D. J. (2008). Adoption of e-government in three Latin American countries: Argentina, Brazil and Mexico. *Telecommunications Policy, 32*(2), 88–100.

Leets, L. (2001). Responses to Internet hate sites: Is speech too free in cyberspace? *Communication Law & Policy, 6*(2), 287–317.

Lessig, L. (1996). The zones of cyberspace. *Stanford Law Review, 48*(5), 1403–1411.

Lessig, L. (2007). In support of network neutrality. *I/S: A Journal of Law and Policy for the Information Society, 3,* 185–196.

Levaggi, R., Orizio, G., Domenighini, S., Bressanelli, M., Schulz, P. J., Zani, C., et al. (2009). Marketing and pricing strategies of online pharmacies. *Health Policy, 92*(2/3), 187–196.

Lipus, T. (2006). International consumer protection: Writing adequate instructions for global audiences. *Journal of Technical Writing & Communication, 36*(1), 75–91.

Lucas, S. E. (1988). The renaissance of American public address: Text and context in rhetorical criticism. *Quarterly Journal of Speech, 74*(2), 241–260.

Lunceford, B., & Lunceford, S. (2008). Meh. The irrelevance of copyright in the public mind. *Northwestern Journal of Technology and Intellectual Property, 7*(1), 33–49.

Lunceford, S. (2008). Network neutrality: A pipe nightmare. *University of Baltimore Intellectual Property Law Journal, 17,* 25–53.

Madison, J. (1865). To W. T. Barry, August 4, 1822. In *Letters and other writings of James Madison* (Vol. 3; pp. 276–281). Philadelphia, PA: J. B. Lippincott.

Marshall, A. M., & Tompsett, B. C. (2005). Identity theft in an online world. *Computer Law & Security Report, 21*(2), 128–137.

McGee, M. C. (1975). In search of "the people": A rhetorical alternative. *Quarterly Journal of Speech, 61*(3), 235–249.

McGee, M. C. (1980). The "ideograph": A link between rhetoric and ideology. *Quarterly Journal of Speech, 66*(1), 1–16.

Morse, E. A. (2007). The Internet gambling conundrum: Extraterritorial impacts of U.S. laws on Internet businesses. *Computer Law & Security Report, 23*(6), 529–536.

Mwesige, P. G. (2004). Cyber elites: A survey of Internet café users in Uganda. *Telematics and Informatics, 21*(1), 83–101.

Nosko, A., Wood, E., & Desmarais, S. (2007). Unsolicited online sexual material: What affects our attitudes and likelihood to search for more? *Canadian Journal of Human Sexuality, 16*(½), 1–10.

Nunziato, D. C. (2003). Freedom of expression, democratic norms, and Internet governance. *Emory Law Journal, 52*(1), 187–279.

Ojedokun, A. A. (2005). The evolving sophistication of Internet abuses in Africa. *The International Information & Library Review, 37*(1), 11–17.

Oriola, T. A. (2005). Advance fee fraud on the Internet: Nigeria's regulatory response. *Computer Law & Security Report, 21*(3), 237–248.

Peitz, M., & Waelbroeck, P. (2006). Piracy of digital products: A critical review of the theoretical literature. *Information Economics and Policy, 18*(4), 449–476.

Poindexter, J., Earp, J., & Baumer, D. (2006). An experimental economics approach toward quantifying online privacy choices. *Information Systems Frontiers, 8*(5), 363–374.

Robison, K. K., & Crenshaw, E. M. (2002). Post-industrial transformations and cyberspace: A cross-national analysis of Internet development. *Social Science Research, 31*(3), 334–363.

Sableman, M. (1999). Link law: The emerging law of Internet hyperlinks. *Communication Law & Policy, 4*(4), 557–601.

Salus, P. H. (1995). *Casting the net: From ARPANET to Internet and beyond.* Reading, MA: Addison-Wesley.

Schell, B. H., Martin, M. V., Hung, P. C. K., & Rueda, L. (2007). Cyber child pornography: A review paper of the social and legal issues and remedies—and a proposed technological solution. *Aggression and Violent Behavior, 12*(1), 45–63.

Schiller, H. I., & Miège, B. (1990). Communication of knowledge in an information society. In J. Berleur, A. Clement, R. Sizer, & D. Whitehouse (Eds.), *The information society: Evolving landscapes* (pp. 161–167). New York, NY: Springer-Verlag.

Seay, J. E. (2008). Hang 'em high': Will the Recording Industry Association of America's new plan to posse up with Internet service providers in the fight against online music piracy finally tame the wild Internet? *Journal of Intellectual Property, 16,* 269–296.

Shrimali, G. (2008). Surplus extraction by network providers: Implications for net neutrality and innovation. *Telecommunications Policy, 32*(8), 545–558.

Simpson, P. (2003). Spoofed identities: Virus, spam or scam? *Computer Fraud & Security, 2003*(10), 6–8.

Sims, L. J. B. (2003). Mutiny on the net: Ridding P2P pirates of their booty. *Emory Law Journal, 52*(4), 1907–1947.

St.Amant, K. (2005). A prototype theory approach to international website analysis and design. *Technical Communication Quarterly, 14,* 73–91.

Starke-Meyerring, D., Duin, A. H., & Palvetzian, T. (2007). Global partnerships: Positioning technical communication programs in the context of globalization. *Technical Communication Quarterly, 16*(2), 139–174.

Stone, B., & Helft, M. (2009, April 27). One Internet village, divided: In developing countries, web grows without profit. *New York Times,* p. B1.

Sunstein, C. R. (1995). The First Amendment in cyberspace. *The Yale Law Journal, 104*(7), 1757–1804.

Thomas, D. (2002). *Hacker culture.* Minneapolis, MN: University of Minnesota Press.

Thrush, E. A. (1993). Bridging the gaps: Technical communication in an international and multicultural society. *Technical Communication Quarterly, 2,* 271–283.

Tyacke, N., & Higgins, R. (2005). Internet law: The legality of Internet "pop-up" ads. *Computer Law & Security Report, 21*(3), 262–265.

US fraud victims face Internet auction rip-off: FBI hotline gets more than 200,000 complaints. (2007). *Computer Fraud & Security, 2007*(4), 4–6.

Waites, M. (2005). *The age of consent: Young people, sexuality, and citizenship.* New York, NY: Palgrave Macmillan.

Wang, S. S., & Hong, J. (2010). Discourse behind the forbidden realm: Internet surveillance and its implications on China's blogosphere. *Telematics and Informatics, 27*(1), 67–78.

Weaver, R. M. (1953). *The ethics of rhetoric.* Chicago, IL: H. Regnery.

Whorf, B. L. (1956). *Language, thought, and reality: Selected writings* (J. B. Carroll, Ed.). Cambridge, MA: MIT Press.

Whybrow, J., & Reed, C. (2002). Internet law: Will online gambling become lawful? *Computer Law & Security Report, 18*(1), 53–55.

Williams, K. S. (2003). Controlling Internet child pornography and protecting the child. *Information & Communications Technology Law, 12*(1), 3–24.

Winner, L. (1977). *Autonomous technology: Technics-out-of-control as a theme in political thought.* Cambridge, MA: MIT Press.

Winseck, D. (2008). Information operations "blowback": Communication, propaganda and surveillance in the global war on terrorism. *International Communication Gazette, 70*(6), 419–441.

Wong, R., & Garrie, D. B. (2007). Network neutrality: Laissez-faire approach or not. *Rutgers Computer and Technology Law Journal, 34,* 315–365.

Yang, K. C. C. (2007). A comparative study of Internet regulatory policies in the greater China region: Emerging regulatory models and issues in China, Hong-Kong SAR, and Taiwan. *Telematics and Informatics, 24*(1), 30–40.

Yoo, C. S. (2005a). Beyond network neutrality. *Harvard Journal of Law & Technology, 19*, 1–77.

Yoo, C. S. (2005b). Network neutrality and the economics of congestion. *Georgetown Law Journal, 94,* 1847–1908.

http://dx.doi.org/10.2190/LIIC3

CHAPTER 3

The Rules of the Game: Real Legal and Economic Implications of *Second Life*

Marco Antonio Chávez-Aguayo

Over the last decades, information and communication technologies (ICTs) have had an important effect on people's daily lives. Today, many individuals view being connected to the Internet as something indispensable. Some Internet technologies, such as web browser tools, instant messaging, blogs, and social network applications have acquired a more central role in spheres such as work, business, education, culture, research, public service, leisure, and interpersonal relations.

This chapter examines a particular application: the Internet-based three-dimensional (3-D) virtual world (VW) of *Second Life* (SL). SL is an online environment where users located across the globe can connect, interact, and exchange goods and services created by them in this virtual environment, and it has proven to have applications and advantages that benefit many different fields. However, as a relatively new technology, it also brings with it numerous challenges to consider and to analyze. This chapter will review these aspects as well as different economic and regulatory issues affecting and affected by SL in a global context. Due to the novelty of SL and global interactions in it, many of these issues have not yet been researched. Interestingly, such issues have also become more complicated, for they increasingly involve a context wherein people from different nations interact together under various legislations.

LAWS, GAMES, AND REALITY

In his 1988 book *Las Reglas del Derecho y las Reglas de los Juegos* [The Rules of Law and the Rules of Games], Robles Morchón used the metaphor of game rules—especially the kind of games where people directly interact, such as board games—in an educational way in order to facilitate the understanding of the roles different regulations play in a given legal framework. Therefore, within the context of this chapter, taking Robles Morchón's metaphor in the opposite way, the rules of law will be used here to analyze the rules of a "game" (a game-like environment), focusing in two central items:

- The legal and economic implications that a metaphoric universe, such as SL, has in real life (RL).
- The serious issues that a game-like application, such as SL, can have.

Doubtlessly, the features of this VW (interactivity, online connection, persistence, intellectual property exploitation, economic trade and exchange, etc.) have developed an environment that is very similar to RL, but changing for the better some of its aspects. As a result, for instance, people's avatars—the user's graphic representation—mostly appear with idealized body shapes and faces (always young, good looking, and slim, according to occidental codes) and have a high and complex level of interaction.

Laws, however, frequently develop more slowly than ICTs, and this gap between law and technology often means gaps in the regulations affecting the uses of such technologies. The result of such gaps might, in turn, become the source of rights violations, unfairness, and injustice. Moreover, cyberspace has many undetected lawless lands where a lack of global rules could open the door to illegal and damaging behavior. Thus, more research into legal areas involving global interactions across the Internet could help governments, law professionals, developers, scholars, researchers, and users devise new regulatory solutions that could facilitate a more secure and fairer integration of ICTs into our daily lives.

Additionally, while fascination with and use of virtual environments has increased in recent years, there is still a lack of research literature that examines SL and VWs in global contexts. Rather, most of the print publications and Internet posts reviewed by the author were dedicated more to describing the features and the novelties of a given VW than providing a critical or an analytical review of such technologies. Similarly, there appears to be a lack of work that crucially examines such technology in order to find a better way to develop different applications for business, education, culture, science, communication, and social processes. This chapter attempts to fill this gap by focusing on three important issues involved in balancing law, SL, and RL with the objective of finding a better, safer, and fairer experience *in-world.* Those three issues are virtual economy, user's rights, and governability.

CONTEXTUALIZATION

A Rough Definition

Despite the fact that it is often defined and classified as a *massively multiplayer online role-playing game* (MMORPG), there are reasons to believe that

SL is something more. However, to gain a quick understanding of what SL is, this definition can be an effective starting point for later distinction.

To begin, a *role-playing game* is a game wherein users assume the role of fictitious characters (sometimes fantastic, but in SL, the character could also be an ordinary human being) and controls the actions of that character. The graphic representation of the user within the computer application is called an *avatar*. Although, in SL, avatars have not preset characteristics or personality. All these can be customized by the user, who decides which RL features to transmit to the avatar and which ones to make up.

Within this context, the phrase *massively multiplayer online* means that many individuals located all over the world use the Internet to connect to a common computer program (known as a *client*) on a remote server (which hosts the VW and keeps the information of the users). This arrangement allows persons from many nations to join SL and interact with each other there.

Usually, MMORPG players have to beat enemies, solve problems, carry out tasks, or accomplish goals in order to obtain points and increase levels needed to get to more complex stages of the game (the MMORPG). Players then continue with these processes until they achieve the main objective and reach the end of the game. Interaction with other players in MMORPGs allows the kind of teamwork that can enhance a player's progress toward this objective. However, the environment, the script, the story, the obstacles, and the goals individuals encounter in MMORPGs are all previously established by the developer who created that virtual environment.

In SL, all of these factors are different. There are no default tasks or goals for individuals to pursue. Similarly, there is no game script nor are there levels of play or a game end. In fact, most of the content (landscapes, objects, figures, etc.) found in SL is built by the users themselves, who are constantly creating new things. Linden Research, Inc., known as *Linden Lab* (LL), is the service and structure provider (servers, software, websites, staff) that gives SL users (called *residents*) the ability to create various objects used to fill in the *metaverse*, or the online location where SL interactions take place. In this virtual environment, almost all the aspects of virtual life can be customized by the user.

Numerous features of SL allow users to engage in such processes, and García and Martínez (2008) have summarized the main features of SL as follows:

- Multiuser: this is a result of the participation of many users connected at the same time.
- Socialization: facilitates the communication and interaction among users.
- Interactivity: allows reciprocal interaction with the environment (virtual objects and elements, as well as the users represented by their avatars).
- Immediacy: interaction occurs on a real-time basis.

- Immersion: creates the feeling of being submerged in the 3-D environment.
- Persistence: the platform remains regardless of whether the user is connected or not.

Currency

SL is also an economic zone. That is, SL has a kind of internal currency known as the Linden dollar (L$), which can be exchanged for RL money such as U.S. dollars (US$) or other denominations. Within the SL environment, individuals exchange these L$ for a variety of virtual goods and services. There are also many ways to exchange RL money for L$ and vice versa. As a result, many enterprises have arisen to tap this burgeoning market. One of these, LindeX Exchange, is the LL official currency trade service. It offers customers a floating currency rate, operates under "market rules" (although the veracity of this will be discussed later), and is integrated to an individual's SL account. LindeX Exchange allows individuals to use local currency for the purchase/sale of L$, and these online purchases are made via credit card or PayPal. Other companies accept different modes of payment including short message services (SMS), local bank accounts, or international transfers.

Growth

SL is neither the only VW nor is it the oldest. It was released in 2003 when many VWs were already on the market. Nevertheless, SL has been one of the most popular VWs and is perhaps the best known worldwide. It has the world's largest and most diverse community, and during its first years of existence it experienced an exponential international rate of growth. According to LL statistics, the number of SL accounts reached a million in October of 2006 (P. Linden, 2006), and by May 2010, there were more than 19 million registered signups (T. Linden, 2010). Other statistics (as LL stopped publicizing quarterly reports in late 2011) indicated over 38 million created accounts by December 2013 (Nino, 2012, 2013).

However, daily concurrent users (the number of daily users online at the same time) started decreasing since 2009. In May 2009, the average of concurrent users was 62,000 while in May 2010 it was 54,000 (Nino, 2009, 2010) Additionally, median user concurrency from September 2012 to September 2013 was around 43,000 (Nino, 2013).

Recognition of Intellectual Property

In the beginning, SL was not much different from other online VWs nor did it stand out from any other VW for its graphics or its performance. During this time, LL's income came from the payment of a membership fee and a daily tax charged for the use of the objects placed (i.e., created) within the SL environment. Some months after its release in 2003, LL announced “a significant breakthrough” in VW business, and this breakthrough involved recognizing the ownership of the content created by residents (SL users) in-world (while in SL) and allowing those users to retain “full intellectual property protection for the digital content they create” (Linden Lab, 2003). Soon after this announcement, LL’s source of income shifted from the payment of membership fees to virtual land revenues. A free membership was opened, and LL reserved some exclusive benefits for paying members (e.g., being able to buy land directly from LL). This change meant that when a user created something in SL, that user could decide to keep, share, or commercially exploit the content he or she made. That user can also buy virtual land to build on, have a house or a business on, admit or deny the entrance of other residents to, or resell and speculate on something else.

The feature that has perhaps generated the most business is the ability of each resident to create SL content from simple geometric objects (called primitives or *prims*) that can be modeled, sculpted, textured, colored, and combined to create new complex elements. Moreover, these new items can include a programming code (or *script*) that can make the object behave in a given way or carry out some specific activity. All of these elements (textures, modeled objects, programming codes, and their combinations) can be exchanged and traded among SL residents through economic transactions, and this factor has generated an internal trade in such goods. Likewise, there is a thriving service market in SL, which includes sectors for different kinds of virtual jobs including everything from sale agents, managers, consultants, and staff for nightclubs or real estate enterprises to models and escorts.

“SL isn’t a Game”

Due to these features, SL founder and former LL CEO Philip Rosedale (also known in-world as Philip Linden) has repeatedly asserted that, unlike MMORPGs, SL is not a game but “a new version of the world” (P. Linden, 2006). Maybe SL cannot be defined solely as a game; however, it indeed also has an aspect that is so entertaining that it cannot be ignored.

The popularity of SL and its attractive features have caused many companies and enterprises to have an interest in breaking into this VW. Some, for example, have opened a virtual branch office that engages in advertising RL items or selling products for virtual life. And in certain cases, organizations actually have employees working in customer care services in their SL franchise. There are also institutions, organizations, and even governments with virtual headquarters in SL as well as also many 100% virtual businesses, rather small or large, that depend on the activity and wealth of SL to subsist.

THE VIRTUAL ECONOMY

On a platform in which the environment has a videogame appearance, it is easy to think that economic trade and business are just part of the game and that SL activities have no serious economic implications in real-life matters. Beller (2007), however, recommends that economists familiarize themselves with the complexity of a virtual economy and not to confuse the tasks and methods of RL with those applicable to VWs. In their turn, Salomon and Soudoplatoff (2010) reflect on differences between tangible and intangible economies. The two say that virtual economies deserve serious study because of their implications as test beds for economic theory, consumer behavior, and intellectual property.

In terms of Second Life, LL stated repeatedly that they have a hands-off policy in regard to virtual economy and "rule SL with Adam Smith's 'invisible hand' only" (Living a Second Life, 2006). Yet many authors have found signs of interventionism, which certainly affects the development of an internal economy. Because there is still no global regulation of VWs' economy, nothing prevents LL from acting arbitrarily against any interest different from its own. This factor could, in turn, have interesting implications for the development of an international VWs' market.

Definition of L$

The common use of L$ within and outside of SL is very different from the legal definition given by LL. In its Terms of Service (TOS)—the adhesion contract users have to accept electronically in order to gain or keep access to SL—LL gives the following definition:

> Each Linden Dollar is a virtual token representing contractual permission from Linden Lab to access features of the Service. Linden Dollars are available for Purchase or distribution at Linden Lab's discretion, and are not redeemable for monetary value from Linden Lab. (Linden Lab, 2013b)

Through this approach, LL clearly avoids all responsibility toward its customers, reserving the ultimate ownership of the license for itself. LL also reserves the right to revoke the L$ license "at any time without notice, refund or compensation" (Linden Lab, 2013b). Ballen (2008) compares this evasion of defining L$ as money with the "dubious accounting practice" that has been involved in different corporate scandals. (In these cases, financial statements were technically legal, but painted a picture that did not reflect reality.) Thus, the legal definition of a pseudo-currency (used in SL as a mean of exchange for goods and services and as a basis for a trade and commerce system) is not a currency at all, or the property of anyone but LL. So far, this economic approach seems unstable and unsafe for international investors and for those who have a business in SL. It is hard to think of a strong economy based on these practices.

In effect, any economy based on this approach could completely and instantly disappear if LL unilaterally decided to end this financial service, and such a decision would leave all of the VW investors with empty credits and worthless tokens. Such actions, moreover, would be of no consequence to LL, as its users previously agreed that their L$ had no monetary value, and accounts are managed at LL's sole discretion. Unless there were to be some external rule governing LL actions and for the protection of their users, there is nothing to prevent such a situation from happening.

Inflation

The manipulation of the L$ supply in order to show a wealthy economy and artificial growth is evidence of LL's intervention in the virtual economy of SL. However, such actions are generating inflation and weakening the economy of this VW. For example, unlike with free accounts, premium accounts require residents to pay a monthly fee in order to gain some advantages. Such advantages include buying land directly from LL or having wider support service and getting a monthly stipend in L$. To give these stipends to premium accounts, LL has been creating the money from nothing—no economic production is involved and there is no backup of its reserves. Such an approach can lead to interesting financial situations in the global marketplace.

Beller (2007), for example, argues that the income from premium accounts during September 2005 was about US$98,000 (equivalent to L$26 million). The creation of L$ during the same period, however, was about 36 million. This situation is equivalent to increasing the deficit budget by 6% a month—a situation that, when annualized, is almost double that amount. In this context, even if the total amount of the revenue were used to back up the L$ supply, the system is still in deficit. In fact, a considerable deficit should exist in SL, and LL should try to use the revenue it receives to address the regular expenses of the company.

Interestingly, LL has varied the amount of the L$ stipend received by premium residents at different times, and the company has also canceled and restored such stipends. Despite the creation of L$, the currency rate on LindeX Exchange since its creation has been stable all the time, fluctuating a little from L$260 to L$270 per US$1 (Linden Lab, 2013a).

Because the main revenue for LL shifted from premium accounts to the sale of virtual land, the information LL used to publish until late 2011 about its accountability did not provide enough information to know if the total L$ supply—which was 7.4 billion for May 2010 (Linden Lab, 2010a)—is tied 100% to real-world commodities or if there is still a large deficit. The situation worsened since LL stopped reporting its economic situation in 2011. Due to this lack of transparency, there could be a sort of global financial crisis in SL, similarly as it is actually in RL. In fact, there could even be a virtual international bankruptcy should SL residents attempt to change all of their L$ deposits into real money and LL does not have enough backup funds to pay all of the resulting debt. Additionally, because LL is not obliged by its TOS to pay anyone if this kind of crisis occurs, the global users in SL will be very affected by this financial situation. Again, a legal loophole in international regulation leaves the customers in the worst position. Moreover, the involvement of people from many countries makes this situation even more problematic, since there is no national or international institution that looks after the interests of VWs' users and customers or controls VWs' currencies.

Gambling and Bank Prohibition

Another sign of interventionism was the wagering prohibition and the banning of virtual banks in SL. In April 2007, LL invited a number of U.S. FBI officials and U.S. attorneys to enter SL and assess the legality of casinos and in-world wagering. Interestingly, LL later decided to ban gambling in SL, and it did so unilaterally by removing all wagering-related objects and by suspending the accounts of some of the residents involved in wagering activity. Moreover, LL made this move without giving a refund given to the international clients affected by these decisions (R. Linden, 2007; Pasick, 2007; Reuters, 2007).

As a result of this unilateral move, people began to withdraw their L$ deposits made in virtual banks, most of which had an ATM located in the casinos. This massive withdrawal resulted in one large in-world bank called Ginko Financial declaring its insolvency and leaving more than L$200 million (US$750,000) unpaid to customers around the globe. Of course, there was no support for the defrauded customers who discovered that Ginko was operating under such a Ponzi scheme. Furthermore, there was no authority in any nation that these individuals could turn to for a legal remedy to this situation.

A month after the decision to ban wagering was made, LL tried to explain this move by posting a statement saying,

> The most important point is that real-world banks are regulated by real-world laws. Linden Lab does not intend to recreate or subvert real-world laws in any way. We are not aware of any institutions in Second Life that are insured by the Federal Deposit Insurance Corporation or similar governmental agencies in other countries. (Z. Linden, 2007)

However, in January of 2008, LL finally decided to intervene in the virtual economy and prohibited any company that offered interest on any other kind of investment or return. (Some virtual banks, for example, promised an annual interest of up to 60%, and Ginko offered rates of up to 40%.) To support this decision, LL noted it would remove any object that would facilitate in-world banking (K. Linden, 2008).

These measures caused mixed reactions across SL users. Some individuals welcomed the new policy as a way to prevent various unethical practices, for many people had been clamoring for RL regulation, and the ban on wagering represented an attempt by LL to regulate the economy. Yet others saw this move as unwarranted regulatory interference and an imposition on personal freedom.

Of course, these regulatory measures were too weak to balance and strengthen the in-world economy. In this situation, LL could not keep its "free market" rhetoric speech and had to react when they had no other choice. To do so, it stepped into the middle of an absolutely (and wild) free economy, presenting some limited regulation. Control over a virtual economy cannot, however, consist of just eliminating practices that could destabilize the economy or LL business. Instead of isolated measures, systematic control needs to be established. This approach might also help protect customers against fraud or abusive behavior as well as promote a fair and equal market in a VW.

Taxes

Each month, millions of transactions take place within SL, and these transactions occur among SL users as well as between users and LL. Because SL connects people from all around the world, in many cases, money is traveling from one country to another and is being exchanged for various goods and services. If these transactions happened in RL, most of them would be subject to different taxes and duties. But of course nobody is paying taxes of any kind to any government for trade in SL. In fact, the legal limbo in which SL exists makes it an informal economy that is exempt from RL taxes. This world is thus a vast field of tax evasion.

In this legal grey area, many things are taking place. Krangel (2008b), for example, reports that music labels were selling tax-free music (a real-world good) in IMVU, a VW similar to SL. These transactions were done through a service called IMVU Music, a system by which users could buy this music in IMVU and export it outside the real world (e.g., into an iPod or an MP3 player) with IMVU's local currency (IMVU, 2008). LL has no similar service, but all in-word commercial exchange is tax free. Since 2008, a value-added tax (VAT) has been collected in SL, but this tax applies exclusively to European Union (EU) residents and is for membership fees, land purchases, land maintenance fees, and LindeX Exchange transaction fees paid in real-world currency. The criteria LL uses to determine who to charge EU VAT is a matter of the user declaring his or her country of residence. In this system, there is no further location verification by IP address despite the fact that LL stated it will do so (Linden Lab, 2010b). Thus, for SL users to evade this tax, all they need to do is claim that they reside in a country outside of the EU.

In 2005, the *New York Times* reported on an industry in China that employs young gamers who spend a lot of time playing MMORPGs in order to acquire virtual items that can later be sold online for real money. And, as happens in a VW, there are no taxes on the sales of these items. The scope of such activities, moreover, is sizable. One source, for example, estimates that over 100,000 Chinese individuals were working full-time in this "gold farmer" business (Barboza, 2005).

The global financial scale of this situation has apparently come to the attention of certain governments. For example, in 2008, China decided to impose a 20% tax on the proceeds of this sale of virtual goods (Krangel, 2008a). The same year, the Internal Revenue Service (IRS)—the tax collection agency of the U.S. federal government—studied the emerging issues arising from VWs. The purpose of this study was to determine what in-world activity could be taxed, what procedures could be used, and what difficulties might be encountered in taxing VW transactions (IRS, 2009). Regardless of some measures that have been taken in certain countries, there is still no coordinated global action to regulate taxes in virtual worlds.

THE PROTECTION OF USER'S RIGHTS

As seen before, the relationship between customers and LL is not equitable. To gain access to the service provided by LL, the user must accept LL's TOS (an adhesion contract) and do so without any negotiation. (It is a "take it or leave it" situation.)

According to the TOS, LL has no responsibility or liability over many situations such as:

- Accepting returns or providing refunds for services purchased from LL.
- Changing or eliminating features or contents of the service.
- Monitoring the service, conduct, or content of third parties or other users.
- Interruption of service and loss of server data.
- Managing, regulating, controlling, and/or modifying or revoking the L$ license without notice, refund, or compensation.
- Revoking the Virtual Land license rights without notice, refund, or compensation.
- Investigating damages or disgorgement or comparable equitable remedy for lost data or lost profits.
- Suspending or terminating a user's account if they determine at their discretion that it is necessary or advisable to protect the rights or interests of LL, the SL community, or any third party.
- Suspending or discontinuing the service, in whole or in part, for any reason without compensation.
- Change the TOS, unilaterally by simply notifying the user 30 days in advance. (If the user does not accept the change, he or she cannot gain access to the service again, not even to take out their holdings, such as L$ balance or objects. Only the US$ account balance can be refunded). (Linden Lab, 2013b)

These TOS were modified in May 2010 to reinforce the position and power of LL after it was sued for the second time by residents who alleged their property rights had been taken from them.

The first lawsuit was *Bragg v. Linden Lab and Philip Rosedale*, and it began in 2006. In this instance, Marc Bragg, a lawyer, found a way to buy land in an SL auction and to do so at below-market prices. LL alleged such actions constituted a violation of the TOS, and it canceled Bragg's account. Bragg sued LL, arguing that, due to this action, he also lost virtual assets valued at between US$4,000 and US$6,000. After various filings and motions, the civil action was settled before a final legal verdict could be handed down as both parties agreed to a confidential settlement (Duranske, 2007; Nino, 2007). However, the U.S. District Court overseeing this case did decide on issues that could be of relevance to future cases:

1. The mandatory arbitration provision in the then current TOS was considered unenforceable.
2. Interaction with a person in a virtual world can satisfy a state's "minimum contact" requirement for personal jurisdiction (i.e., about including Philip Rosedale in the lawsuit due to his nationwide personal marketing efforts to publicize SL). (Robreno, 2007)

The second lawsuit, *Evans et al. v. Linden Lab and Philip Rosedale*, filed in April of 2010, was a class-action suit filed by Jason Archinaco on behalf of virtual landowners. In the suit, Archinaco alleged that LL misled players into thinking that they owned their virtual lands and was suing LL for fraud and damages in excess of $5 million. In general terms, the virtual landowners complained about consumers' rights and how LL enticed them to purchase virtual property by promising ownership of that property provided the purchaser agree to the terms of LL's TOS. According to the virtual landowners, once purchases were complete, LL unilaterally changed the adhesion contract and thus prevented consumers from accessing their online holdings unless the consumer agreed to the new terms of use (Archinaco, 2010). To support these claims, the wronged parties repeatedly cited the marketing campaign Philip Rosedale used in order to draw them into this initial agreement. Rosedale, after all, was CEO of LL at the time, and he repeatedly promised SL residents that the items they bought in-world (e.g., virtual land) were theirs to own. A key document in this case was a 2003 LL press release announcing the recognition of intellectual property rights: "The preservation of users' property rights is a necessary step toward the emergence of genuinely real online worlds" (Linden Lab, 2003).

The lawsuit also stated SL's motto at that time, appearing at the frontpage of its website, which described it as being a VW "imagined, created and owned by its residents" (Archinaco, 2010). However, when this second lawsuit was presented, all this ownership discourse was removed from SL's website and advertising. In fact, in the TOS of May 2010, no mention is made of any kind of ownership by individuals in SL.

Additionally, current LL's TOS requires SL residents to "acknowledge that Virtual Land is a limited license right and is not a real property right or actual real estate, and it is not redeemable for any sum of money from Linden Lab" (Linden Lab, 2013b). LL's policies thus show very weak or little (if any) protection of users' rights. In fact, the unilateral ability of LL to change the TOS, decrease users' rights, and increase the authority of LL are central factors in gaining advantage and acting as a monopoly.

This case reveals that there is still the pending task of trying to more accurately reinforce and safeguard the rights of VWs users from different countries. LL's TOS says that no matter where the users are registering or connecting from, the contractual relationship between the users and LL is established with LL's headquarters in San Francisco, California. So, unless the user is within the United States, any dispute involving SL must be solved according to the legal frameworks of different countries.

GOVERNABILITY

LL is, after all, a company that pursues profit and is pressed by global markets. It is therefore understandable that the organization's priority is to secure the profit-making of the business rather than building a fair and equal platform. Interestingly, LL has stated and flaunted its hands-off policy with regard to the economy and to user-to-user relations several times. Yet the company has repeatedly acted when its own interests were endangered, and it did more to prevent specific situations from happening than to opt for systematic control or intervention. As it was discussed before in this chapter, LL acted against bank speculation when such speculation threatened the confidence or the stability of SL's economic system. The company also intervened in online wagering in order to avoid a legal issue with U.S. federal legislation—the same thing LL did when it started (hypothetically) collecting the VAT for EU residents in order to follow EU laws. The discretion of LL's decisions, creating L$, inflating financial assets in SL, changing the TOS to reduce LL's liability, eliminating references to user's property or ownership, and so on are better understood in terms of LL looking after its sole interests.

Philip Rosedale repeatedly defined SL as a nation, or even "a new version of the world" (P. Linden, 2006). Some commentators, such as Beller (2007), have even compared LL to a small foreign country that depends on tourism for its economy. However, given the way in which LL governs the VW, critics such as Archinaco (2010) have described SL as a dictatorship. This assessment is based on the fact that LL causes investors to build an infrastructure and does so by saying that the individual user has "ownership" rights. Yet the underlying intention of LL seems to be to "nationalize" assets and suspend investor accounts.

For these reasons, individuals like Grimmelmann (2009) note that SL is better defined as a feudal society. That is, in SL, LL acts as the lord/developer that controls property, maintains deed records, and collects taxes. If, however, the user doesn't pay the monthly tiers or behaves in an improper way (as defined by LL), then LL freely seizes the user's property and suspends the user's account. LL does so based on a "law" established unilaterally by LL (the TOS) and that gives LL total permission to engage in such actions without any consequences.

It is actions such as these that prompt scholars and users to clamor for a more democratic experience in VWs. Some alternative models can be developed to achieve this intention. A representative body could, for example, govern a VWs by looking after user's interests rather than the sole interests of the profit-seeking company that owns a given platform. Ideally, RL governments should not excessively regulate VWs. Rather, common law may be used instead of regulation, and the development and growth of VW communities should be encouraged by promoting the standardization and transparency of agreed rules (Fairfield, 2008; Jakobs, 2009; Zwart, 2009).

Decline of SL

SL has had ups and downs throughout its history. At its start, people were surprised with this novelty of ICTs and were curious to discover and experience its many features. As Masnick (2008) notes, a few years ago "you could hardly go a day without hearing some press story hyping up how the virtual world Second Life was the biggest thing since the Internet itself." During this time, many RL enterprises came into SL both to advertise and to sell products to different markets. Additionally, politicians began campaigns in SL, and institutions, universities, and even embassies were established throughout the in-world. (Reuters news agency even had correspondents reporting from within SL.)

However, the bubble started deflating. After the bank prohibition, or better said, the intervention on SL economy, there were many signs of economic recession: negative growth, fewer user-to-user transactions, and lower user profits (Holyoke, 2008). Maybe these were some of the reasons for LL to stop releasing official reports about its economy and growth in late 2011. In fact, as Gartner (2008) explains, some 90% of all business forays into VWs fail within 18 months because organizations focus on technology rather than on understanding the needs of the employees using it. "Further reasons for the high failure rate include starting projects for the 'cool' factor or because competitors are doing it" (Gartner, 2008). Reuters eventually closed its SL bureau in 2008 and pulled its full-time SL reporter, Eric Krangel, who later suggested that SL must:

- Build more content for newcomers
- Acknowledge that its reputation is now a liability
- Simplify the user interface
- Abandon the idea that it is a business application

Krangel also adds that lately, hanging out in the virtual world was "like watching paint dry" (Krangel, 2008c). In fact, in 2009, the U.S. Better Business Bureau gave LL an "F" grade due to the number of complaints filed against its business, to its failure to respond to complaints, and the length of time LL has taken to resolve complaints (Krangel, 2009).

Alternatives to SL

The recent history of ICTs shows that once people learn to use a particular type of application, they can easily migrate to a similar kind of application designed by a different developer. The case is all too common in many aspects of our regular lives: the move from Netscape to Explorer to Mozilla to Chrome; the move from Altavista to Yahoo to Google, and so on. In this environment, competition is stiff, and the organization that offers better performance or service will attract the most consumers.

The actual features of SL in some way make it very particular: intellectual property recognition, functioning trade system, large user community, high applicability, the name is well known worldwide because of all the publicity made on the news, and so on. As a result, it would be hard to find direct competitors. Nevertheless, there are initiatives, such as OpenSimulator (or *OpenSim*), which is an open-source application server for hosting VWs. On this open-source platform are many individuals and organizations running different VWs (with the same appearance as SL) that can be interconnected. Despite looking like a good alternative to SL, these VWs still represent an unstable world (most of the platforms are not online 24/7). Moreover, it is never crowded at all nor does it have an economic framework yet. It is a place wherein there is still no real "authority" or agency that controls or coordinates these servers in order to standardize them and make them offer a regular and stable service, particularly as most of them are nonprofit initiatives and thus have fewer resources. However, this situation could represent the next step in the evolution of VWs and related attempts to regulate them.

CONCLUSION

VWs indeed have many interesting features with great applicability in different spheres. For LL, the creation of SL has been a great achievement for which they must be recognized. However, because LL does not seem eager to develop this environment in a fairer and more equitable way, it is hard to think that the future of VWs might be in the hands of a profit-making enterprise.

More involvement is needed from governments around the world. Governments, however, should not try to overregulate virtual platforms. Rather, they should try to become more aware of them in order to understand their logic and establish a coordinated and systematic legal framework for improving and enhancing the development of VWs and its capabilities and features. Users might also be able to self-govern a platform and hence make it a more democratic application.

A globally decentralized network of VWs with a standardized and controlled economy and a representative authority might be the next stage to achieve. This wish is probably not as utopian as it seems. Only 20 years ago, most of the technology and advances that are part of our everyday lives today would have sounded equally idealistic. Today, they are reality.

The implications and applications of VWs still sound very illusory and irrelevant to many people, be they users, potential users, scholars, or professionals. A better understanding and integration of ICTs into daily life is becoming increasingly necessary in order to keep track of the information society. Hopefully, a time will come when ICTs help to bring together people around the world instead of emphasizing division and inequality. It seems like there is a lot of work to do, but at least there is a foundation upon which such work can begin.

REFERENCES AND BIBLIOGRAPHY

Archinaco, J. (2010). *Carl Evans, Donald Spencer, Valerie Spencer, Cindy Carter, individuals, on behalf of themselves and for the benefit of all with the common or general interests, any persons injured, and all others similarly situated, plaintiffs, v. Linden Research, Inc., a corporation, and Philip Rosedale, an individual, defendants.* Retrieved from http://www.virtuallanddispute.com/pleadings/evan_spencer_carter_v_linden_labs_virtual_land_property_rights_class_action_lawsuit.pdf

Au, W. (2008). *The making of Second Life.* New York, NY: Collins.

Barboza, D. (2005, December 9). Ogre to slay? Outsource it to Chinese. *New York Times.* Retrieved from http://www.nytimes.com/2005/12/09/technology/09gaming.html

Beller, M. (2007). The coming Second Life business cycle. *Ludwig von Mises Institute.* Retrieved from http://mises.org/daily/2640

Beller, M. (2008). Wildcat banking in the virtual frontier. *Ludwig von Mises Institute.* Retrieved from http://mises.org/daily/2862

Carr, P., & Pond, G. (2007). *Second Life.* Barcelona, Spain: Random House Mondadori.

Duncan, G. (2007, July 27). Second Life bans gambling. *Digital Trends.* Retrieved from http://www.digitaltrends.com/gaming/second-life-bans-gambling/

Duranske, B. (2007, October 4). Bragg v. Linden Lab—Confidential settlement reached; 'Marc Woebegone' back in Second Life. *Virtually Blind.* Retrieved from http://virtuallyblind.com/2007/10/04/bragg-linden-lab-settlement/

Fairfield, J. (2008). Anti-social contracts: The contractual governance of virtual worlds. *McGill Law Journal, 53.* Retrieved from http://ssrn.com/abstract=1002997

García, A., & Martínez, R. (2008). *Exprime Second Life*. Madrid, Spain: Anaya Multimedia.

Gardiner, B. (2007, August 8). Bank failure in Second Life leads to calls for regulation. *Wired*. Retrieved from http://www.wired.com/gaming/virtualworlds/news/2007/08/virtual_bank

Gartner, B. (2008, May 15). *Gartner* says 90 per cent of corporate virtual world projects fail within 18 months. Retrieved from http://www.gartner.com/it/page.jsp?id=670507

Grimmelmann, J. (2009). Virtual world feudalism. *The Yale Law Journal*. Retrieved from http://yalelawjournal.org/the-yale-law-journal-pocket-part/property-law/virtual-world-feudalism/

Holyoke, J. (2008, February 14). Second Life economy is in a recession. *The Alphaville Herald*. Retrieved from http://alphavilleherald.com/2008/02/second-life-eco.html

IMVU. (2008, November 11). IMVU launches world's first truly social music service. *imvu*. Retrieved from http://www.imvu.com/about/press_releases/press_release_20081111_1.php

Internal Revenue Service (IRS). (2008). *National taxpayer advocate's 2008 annual report to Congress Volume 1*. Internal Revenue Service. Retrieved from http://www.irs.gov/pub/irs-utl/08_tas_arc_intro_toc_msp.pdf

Jakobs, K. (2009). Real standards for virtual worlds: Why and how? *Journal of Virtual Worlds Research, 2*(3). Retrieved from https://journals.tdl.org/jvwr/article/view/717/517

Keen, A. (2006, July 20). Philip Rosedale. *AfterTV*. Retrieved from http://andrewkeen.typepad.com/aftertv/2006/07/interview_with_.html

Krangel, E. (2008a, November 3). Chinese government gets in on gold farming action with new 20% tax. *Business Insider*. Retrieved from http://www.businessinsider.com/2008/11/chinese-government-gets-in-on-gold-farming-action-with-new-20-tax-

Krangel, E. (2008b, November 11). Real-world music comes to fake-world IMVU. Will real-world taxes be far behind? *Business Insider*. Retrieved from http://www.businessinsider.com/2008/11/real-world-music-comes-to-fake-world-imvu-will-real-world-taxes-be-far-behind-

Krangel, E. (2008c, November 21). Exclusive: Why Reuters left Second Life, and how Linden Lab can fix it. *Business Insider*. Retrieved from http://www.businessinsider.com/2008/11/why-reuters-left-second-life-and-how-linden-lab-can-fix-it

Krangel, E. (2009, February 24). Better business bureau gives Second Life an "F." *Business Insider*. Retrieved from http://www.businessinsider.com/better-business-bureau-gives-second-life-an-f-2009-2

Lessig, L. (1999). *Code and other laws of cyberspace*. New York, NY: Basic.

Lessig, L. (2004). *Free culture: How big media uses technology and the law to lock down culture and control creativity*. New York, NY: Penguin.

Linden, K. (2008, January 8). *New policy regarding in-world "banks."* Retrieved from http://community.secondlife.com/t5/Features/New-Policy-Regarding-In-World-Banks/ba-p/605657

Linden, P. (2006a, April 7). Town hall q. *Second Life* Forums Archive. Retrieved from http://forums-archive.secondlife.com/139/65/98819/1.html

Linden, P. (2006b, October 18). When precisely did we hit 1 million residents? Blog Archive. Retrieved from http://community.secondlife.com/t5/Features/When-precisely-did-we-hit-1-million-Residents/ba-p/530131

Linden, R. (2007, July 25). Wagering in Second Life: New policy. Blog Archive. Retrieved from http://community.secondlife.com/t5/Features/Wagering-in-Second-Life-New-Policy/ba-p/585072

Linden, T. (2010, April 28). Second Life economy hits new all-time high in Q1 2010. Blog Archive. Retrieved from http://community.secondlife.com/t5/Features/Second-Life-Economy-Hits-New-All-Time-High-in-Q1-2010/ba-p/657582

Linden, Z. (2007, August 14). The Second Life economy. Blog Archive. Retrieved from http://community.secondlife.com/t5/Features/The-Second-Life-Economy/ba-p/587442

Linden Lab. (2003, November 14). *Second Life residents to own digital creations.* Retrieved from http://lindenlab.com/pressroom/releases/03_11_14

Linden Lab. (2010a). Total L$ supply. *Second Life.* Retrieved from http://secondlife.com/statistics/economy-data.php

Linden Lab. (2010b). Value added tax. *Second Life.* Retrieved from http://secondlife.com/corporate/vat.php

Linden Lab. (2013a). LindeX™ market data. *Second Life.* Retrieved from http://secondlife.com/my/lindex/market.php

Linden Lab. (2013b). Terms of service. *Linden Lab.* Retrieved from http://lindenlab.com/tos

Living a Second Life. (2006, September 28). *The Economist.* Retrieved from http://www.economist.com/node/7963538

Masnick, M. (2008, October 20). Well, at least Second Life won't need a gov't bailout. *Techdirt.* Retrieved from http://www.techdirt.com/articles/20081017/0229442572.shtm

Nino, T. (2007, January 27). Bragg vs Linden Lab—The story so far. *Second Life Insider.* Retrieved from http://archive.is/jQLrl

Nino, T. (2009, August 30). New graph: Median Second Life user concurrency. *Dwell on it.* Retrieved from http://dwellonit.taterunino.net/2009/08/30/new-graph-median-second-life-user-concurrency/

Nino, T. (2010, September 14). Second Life concurrency finally stabilizing. *Dwell on it.* Retrieved from http://dwellonit.taterunino.net/2010/09/14/second-life-concurrency-finally-stabilising/

Nino, T. (2012, March 1). No more quarterly or annual Second Life economy reports, says Lab. *Dwell on it.* Retrieved from http://dwellonit.taterunino.net/2012/03/01/no-more-quarterly-or-annual-second-life-economy-reports-says-lab/

Nino, T. (2013). Second Life statistical charts. *Dwell on it.* Retrieved from http://dwellonit.taterunino.net/sl-statistical-charts/

Pasick, A. (2007, April 4). FBI checks gambling in Second Life virtual world. *Reuters.* Retrieved from http://www.reuters.com/article/idUSHUN43981820070405

Reuters, A. (2007, July 26). Linden Lab outlaws Second Life gambling. *Reuters.* Retrieved from http://secondlife.reuters.com/stories/2007/07/26/linden-lab-outlaws-second-life-gambling/index.html

Robles Morchón, G. (1988). *Las Reglas del Derecho y las Reglas de los Juegos,* Mexico: Universidad Nacional Autónoma de México.

Robreno, E. (2007). Marc Bragg, plaintiff, v. Linden Research, Inc. and Philip Rosedale, defendants. Memorandum. *United States District Court for the Eastern District of Pennsylvania.* Retrieved from http://www.paed.uscourts.gov/documents/opinions/07D0658P.pdf

Salomon, M., & Soudoplatoff, S. (2010). Why virtual-world economies matter. *Journal of Virtual Worlds Research, 2*(3). Retrieved from https://journals.tdl.org/jvwr/article/view/859/624

Senges, M. (2007). *Second Life.* Barcelona, Spain: Universitat Oberta de Catalunya.

Simonite, T. (2008). A virtual refuge from recession. *New Scientist.* Retrieved from http://www.newscientist.com/blog/technology/2008/04/virtual-refuge-from-recession.html

Trouble in paradise. (2007). *The Economist.* Retrieved from http://www.economist.com/business-finance/displaystory.cfm?story_id=9661900

Zwart, M. (2009, October). Piracy vs. control: Models of virtual world governance and their impact on player and user experience. *Journal of Virtual Worlds Research, 2*(3). Retrieved May 20, 2010, from https://journals.tdl.org/jvwr/article/view/663/511

http://dx.doi.org/10.2190/LIIC4

CHAPTER 4

Dimensions and Policies of Metrication and Communication

Ronald L. Stone

The process of metrication involves the adoption of the metric system as the primary mechanism for reporting various measurements in a range of contexts. Moreover, metrication is connected to a global context of measurement that is traceable to the International System of Units, which influences a range of technical and scientific activities related to the collection and the reporting of information. As a result of these factors, metrication presupposes a range of legal implications related to conveying accurate and acceptable information in myriad international situations. Such factors are further complicated by online media that project measurements (and the systems used to report numeric information) designed by and for members of one culture into broader, global contexts. Thus, computational technologies create a context that could result in unexpected, international legal problems (e.g., the 1999 loss of the Mars Climate Orbiter spacecraft was, in part, due to miscalculations associated with calculations done in customary expressions of pound-seconds instead of metric units of newton-seconds) (Stephenson et al., 1999). These connections to communication have important implications for technical communicators, particularly those individuals who are working in a global context who are developing materials for broader international audiences or who are using online media to disseminate information globally.

This chapter provides readers with an overview of what metrication is and an examination of its implications as related to technical communication practices. In so doing, the author examines how metrication might influence technical communication practices in global contexts. To achieve this objective, the chapter begins with a summary of certain international and industrial initiatives for the communication of metrication best practices. The chapter then examines the current state of metrication in light of a focus on best practices for metrication as it relates to communication practices. In so doing, the chapter also notes how the global nature of online media can affect such factors.

METRICATION COMMUNICATION AND STANDARDS

Metrication usually refers to the International System of Units (SI), also known as the modern metric system, and it can be defined as a process of learning about or practicing the use of metric measures (AAT, 2009b). The Système International d'Unités (SI) was established by an international treaty, the Convention du Mètre, and it is presently used throughout the world (BIPM, 2006). Although measures are used everywhere in society, they have a particularly important role in science, engineering, commerce, medicine, and travel (Becker et al., 2007). Moreover, the development of an international system of measures supports international cooperation in areas of commerce and science by promoting consistent communication about weights and measures. Some of the topics in this review that might be particularly relevant to scientific and technical communicators include standards for style and usage when writing, in addition to concerns related to designing documents for translation or localization. As many of the industries and sectors in which scientific and technical communicators work are metricated to a large extent (e.g., engineering), it is important that technical communicators become aware of the basic principles and usage conventions for writing about metric scales of measurement.

How else is metrication an important topic for technical communicators? The very definition of things like units of measure can be usefully approached in terms of Aristotelian concepts of first principle (Aristotle, 1991). For writers, the determination of a definition often begins with reference to a selected style guide or reference resource. Yet when analyzing topics of metrication within the framework of a first philosophy, other types of disciplinary (e.g., scientific or technical) knowledge are considered (Gómez-Lobo, 1978). This situation calls for the involvement of other scientific or technical specialists for purposes of development or review.

With skills for developing information about metrication, technical communicators can be more prepared for updating the presentation of data in ways that can add value to users of technical documentation. The use of metric units, which are practicably related in terms of decimal multiples and submultiples, can provide benefits to both developers and end users. Moreover, as online media now make widespread global audiences the inevitable end users for web-delivered materials, metrication can help technical communicators present data in ways that better meet the expectations of a broader audience of international online users. It may thus serve technical communicators well to consider how the presentation of information in terms of best practices for metrication can facilitate the presentation of information in global contexts.

Technical communicators should also beware of how the presentation of information in terms of pre-metric expressions can affect a communication, both at home and abroad. At best, uses of pre-metric (customary, default, or legacy) expressions such as inch, foot, and mile are antiquated. At worst, pre-metric expressions are unclear, incorrect, clichéd, or pseudoscientific.

Although certain aspects of non-metrication may be entrenched in some domains of engineering, or entrenched in some customary practices, technical communicators should also be aware of some of the costs or consequences of non-metrication, particularly in relation to international settings. Costs or consequences can vary depending on the situation. The severity of a consequence can range from mild inconvenience to critical danger to life or property. Some disciplines such as medicine, for example, might depend upon metrication for determining correct dosage of a medication (Juengel, 2011). In such instances, the use of non-metric measures could lead to inadequate or ineffective treatment or even to the death of a patient. Moreover, when shared within a greater, global context via online media, such as a medical information website, displays of unrecognized systems expressions of measurement might render information unusable or even cause damages or death depending on the nature of the data. (Cultures can, for example, have different interpretations of the volume indicated by the measurement of "teaspoon," a factor that can become problematic in terms of distributing the correct dosage of a medication; U.S. Department of Health and Human Services, 2001). Technical communicators might thus take note of some highlights of a history of metrication in order to understand the importance of such differences.

Early Metrication

A history of metrication is complicated, and it predates the imperial, colonial, revolutionary, and constitutional political environments of North America and Europe in the 1700s and 1800s. In fact, topics related to the ethics or morality of measurement, such that a smaller measure should not be presented as a larger measure in trade, appear in the early texts of both Christian and Islamic books (Naughtin, 2008). Interestingly, one of the first English documents to declare constitutional or civil rights, the Magna Carta (established in England, AD 1215), contained guarantees for uniformity of weights and measures.

By the time of the Magna Carta, when Roman numerals were still widely used for writing numbers, at least some European mathematicians were writing about the use of decimal (Arabic) numerals for calculations (Naughtin, 2008). By the time of the European Renaissance, a few centuries later, various practitioners had produced a number of cultural, scientific, and engineering developments in various regions of Europe, and some of the characteristics of a uniform system of measure were being formulated. In 1585, Simon Stevin of Belgium published both Flemish- and French-language editions of books on the use of decimal numbers and decimal fractions in mathematics (Naughtin, 2008).

Subsequently, in 1668, the Royal Society in England published a philosophy text by John Wilkins that included a chapter on measurements. Many of the ideas on measurement presented in this text were later adopted into the metric system (Naughtin, 2007), and in that text, Wilkins also proposed a universal standard measure that could be expressed in terms of decimal multiples and submultiples. He also suggested that the system be established as an international standard that could be communicated regardless of local language—in essence, a system for conveying technical information uniformly across linguistic and cultural barriers (Naughtin, 2007). Through ideas such as these, connections between practices of measurement and methods of international technical communication began to form.

Pat Naughtin's (2007) commentary on Wilkins' work also highlights some points of particular interest to technical communicators. These include a recognition of needs for clear communication among different nations about methods of measurement. The idea is that by creating a mechanism for clear communication across cultural lines, standards of measurement can serve as a guideline technical communicators can use to add value by recognizing and addressing such conversions in a way that facilitates the international sharing of information.

The overarching idea is that the use of common definitions of measure can reduce considerably the time and cost of doing business in other regions. Calendars can also be regarded as a topic of metrication that, for technical communicators, can involve choices of style and formatting in the course of preparing information for translation or localization. For example, the Gregorian calendar first adopted in Italy in AD 1582 during the European Renaissance was subsequently used by many other nations during transitions that occurred from the 1500s to the 1900s (Tondering, 2008). Although most regions have made the transition from the Julian calendar to the Gregorian, some communities (e.g., members of the Eastern Orthodox Church) still observe the Julian calendar for civil or religious use.

An Alliance for the Advancement of Technology, Integrated Chronological Applications System (AAT ICAS) 2062 standard for the usability of date and time information (extending some principles described in ISO 8601) further reflects some contemporary developments in sharing technical information, particularly via online media, with international audiences. This standard comprises an ICAS conformance level A with checkpoints for guidance with reference to Web Content Accessibility Guidelines (WCAG) standards of the Web Accessibility Initiative (WAI) at W3C. AAT ICAS guideline 2062-01 provides guidance for preparing technical date and time information that is clear and usable for web audiences. Further guidelines encompass the normative representation or labeling of date and time scales and thus provide technical communicators with an heuristic they can use when sharing such information with international audiences.

International Metric System

International development of the metric system occurred amidst the revolutions in France and the United States. In 1787, the U.S. Congress adopted a national constitution, granting Congress a power to "coin money . . . and fix the standard of weights and measures" (Article 1, Section 8). The United States also introduced the first decimal currency, a dollar consisting of 100 cents: "fugio cent" coins were minted and entered circulation immediately following their design by Benjamin Franklin in 1787 (Naughtin, 2008). A few years later, in 1795, the Republic of France adopted the French Academy of Sciences recommendation and legally created a Système métrique décimal that is very much like the metric system we use today (Naughtin, 2008). French law included a provisional standard for the metre and a prefix scheme for decimal multiples and submultiples.

At roughly the same time, reviews of a survey by Delambre and Méchain to determine the distance along a line of longitude from the equator to the North Pole were conducted by an international committee of experts in the late 1790s (Naughtin, 2008). These factors prompted France and Belgium to quickly establish the resulting metric system as their official standard. An international commission convened by France in 1872 took preparatory steps toward creating the Metre Treaty (Naughtin, 2008), which was subsequently established with 17 signatories, including the United States, in 1875 (U.S. Metric Association, 2005). Foundations for metrication as a global standard of measure were now in place. International communications concerning measure could now reference a global standard, without any need of puzzling over which local pre-metric expressions were what measures.

Metric Modernization

The processes of metrication or measurement can be subject to a variety of issues concerning the science, practice, or linguistic expressions involved in areas of metrication. Sources of "great difficulty" (in the words of Wilkins, as cited in Naughtin, 2008) about metrication have included the use of customary or pre-metric measures, differences in regional languages, trade protectionism, and processes of scientific development. These issues have sometimes slowed the adoption of a system of measurement that is otherwise more practicable for more types of measure than any customary pre-metric custom of measure. And all of them can affect the efficacy with which technical information is shared with members of other cultures or within global contexts.

Toward keeping a focus on the communication of metrication best practices, I should wish to point out that attention to issues of language for communication, and a consideration of processes of scientific development, can be regarded to concern the development of best practices. In contrast, issues of the use of customary or pre-metric measures, or issues of trade protectionism (insofar as they do not reflect a best practice) can be regarded as hindrances to the communication of metrication best practices. More about these points will be described later in this entry.

For purposes of brevity, I will try to highlight a few selected ideas about the modernization of metrication that may be most relevant to technical communication. In the century following the initial establishment of the Metre Treaty, countries where uses of inch-pound measures remained customary (including Australia, Canada, the United Kingdom, and the United States) prepared initiatives for metrication and increased the use of metric units for a variety of areas of practice. Canada and Australia metricated in the 1970s (Reid, 2008a, 2008b). However, the United States remains only partially metricated.

At any rate, in 1960, the 11th Confèrence Général des Poids et Mesures (CGPM) of the Metre Convention adopted the name Système International d'Unités, with the international abbreviation "SI." (The official English translation of the French name is "International System of Units [SI]", and the official short-form expression of the system, in any language, is "SI"; BIPM, 2006.)

What this means for business in both international and local markets is that technical communicators can reference best practices for metrication communication and thereby improve the quality of documents and user experiences. (By providing international users with a common standard of measurement, one avoids prospective legal disputes linked to miscommunication resulting from converting measurements from one scale to another.) Another implication for technical communicators is that a strong case can be made for metrication communication as an appropriate approach to fulfilling quality and user experience in international settings by providing a common standard for user experiences in a broad range of international contexts. It may further be that there is no clear need for the use of pre-metric expressions of measure in business and no clear dependence on uses of pre-metric expressions for accomplishing quality and user experience. In any case, metrication offers a number of opportunities for technical communicators to add value to global users of information by helping make information more accessible on an international level. More importantly, it greatly reduced the degree of international confusion that could otherwise lead to international legal disputes relating to items such as fulfilling contracts (e.g., the contracted amount purchased was delivered) or meeting regulatory requirements (e.g., the volume capacity or displacement a mechanism is required to meet to be considered "compliant" in certain markets).

Current State of Metrication

The current state of metrication is subject to advances in science and technology that can improve the accuracy and practical realization of measurement. In 2005, the International Committee for Weights and Measures (CIPM) published a recommendation to begin the process of redefining the kilogram, ampere, kelvin, and mole in terms of exactly known values of physical constants that are "universal, permanent, and invariant in time" (Mills, Mohr, Quinn, Taylor, & Williams, 2006). An example of this was the redefinition in 1960 of the metric unit of length, the metre (meter), in terms of a path of light traveling in a vacuum, rather than in terms of a physical prototype that might be subject to change over time (BIPM, 2006). These developments, in turn, create an interesting legal context that affects how information can and should be shared in a number of international contexts.

Various metrication policies that are relevant to legal aspects of technical communicators include the following:

European Community

The European Council directive 80/181/EEC of the European Economic Community (EEC, 2000) designated SI for use by member states. A 1999 allowance for the use of "supplementary indications" such as inches and ounces was scheduled to expire at the end of 2009, yet a recent amendment of the directive to indefinitely permit the use of these supplementary indications eases certain requirements for the exclusive use of metric units on product labeling in Europe. The amendment is also intended to encourage the United States to approve legislation to permit the use of metric-only labeling on products subject to federal regulation, instead of requiring both customary language and metric units on product labeling, without any recourse to trade barriers (EU, 2007). In effect, the legal issue at stake is that of compliance. That is, if metric labeling is required to distribute a product within the EU, then companies that fail to do so could find themselves cut off from one of the world's most lucrative markets due to, essentially, a legal issue.

Also in play is legislation related to translation. That is, according to EU law, the documentation for many kinds of products needs to be translated into the language of EU member states before a product can legally be sold in those markets (Language requirements in Europe, 2012). As a result, items that slow the translation process, such as the translator's need to convert measurements from one expression to another, create a barrier to how quickly products can legally enter these markets.

United States

Recent metrication policy for the United States is summarized in an announcement in the Federal Register and in special publications of the National Institute of Standards and Technology (NIST, 1998). NIST Special Publication 811 (Taylor & Thompson, 2008), "Guide for the Use of the International System of Units (SI)," provides guidance about legal and stylistic matters for authors writing for NIST or other U.S. federal agencies. The usage and style material in Special Publication 811 includes an explanation of American spellings and usage that are used in NIST documents. For example, "meter" and "liter" are used instead of "metre" and "litre." A number of guidelines review the rules for properly writing SI units. Other guidelines concern avoiding usage that might be subject to different interpretations due to regional variation in terms like *million, billion*, and *trillion* (and thus to avoid terms like *ppm, parts per million*). Advice that the term *weight* should only be used unambiguously, to avoid confusing the concepts of mass and force, is explained with the advice that the term *atomic weight* should henceforth be replaced by the term *atomic mass*. These guides, in turn, affect how technical communicators work for U.S. government agencies, for they determine how individuals must document information in a manner that meets government requirements.

BIPM

The Bureau International des Poids et Mesures (BIPM) is an intergovernmental organization that was established by the Metre Convention in 1875, and the organization continues to ensure worldwide coordination of measurement standards. This coordination involves establishing fundamental standards and scales of measure, comparing national and international standards, and ensuring coordination of measurement techniques (BIPM, 2006).

The BIPM website provides access to many documents and resources, including the SI brochure "The International System of Units (SI)" and the journal *Metrologia*. The CIPM, the group that oversees the work of BIPM, recently adopted a recommendation in preparation for redefining the kilogram, ampere, kelvin, and mole so that these units are linked to exactly known values of physically invariant values (Mills et al., 2006). A redefinition of the kilogram in terms of a fundamental physical constant that is "universal, permanent, and invariant in time" would mean that SI would not depend on the prototype kilogram, an actual physical object that may be subject to changes over time. Since 1960, the meter has been defined in terms of the length of a path of light traveling through a vacuum (BIPM, 2006). The unit of time is also defined in terms of a natural atomic frequency, from which the SI second is scaled (Guinot, 1997). A future kilogram defined in terms of a physically invariant value would make possible the definition of all SI units in terms of two fundamental units of time and length (Wignall, 2007).

From a legal perspective, these developments can affect when an organization might need to meet certain documentation requirements in order to be allowed to sell products in a given overseas market. It could also affect facets of liability and of meeting contractual obligations. In the case of liability, the issue becomes the need for a common understanding of weights and measures as they relate to things such as the correct (i.e., safe) uses of a product (e.g., the settings of a particular machine). In such instances, any confusion or miscalculation relating to how to correctly set a mechanism could result in damage to the item or in the injury of individuals using it. In the case of contracts, the issue would be that different interpretations of measurements mean that contracting parties have different expectations of if and when the measurement-related conditions of a contract have been met (and even if misrepresentation or fraud might be suspected).

IEEE

The Institute of Electrical and Electronics Engineers (IEEE) has implemented a metrication policy (IEEE, 2007) for the exclusive use of metric units in the normative (prescriptive) portions of new and revised standards. The implementation of this policy provides an allowance for parenthetical or supplemental use of inch-pound data "after the metric unit if the sponsor believes that the audience for the document would benefit from the inclusion of inch-pound data, based on concerns for safety or clarity" (IEEE, 2007, p. 19).

The policy also provides for the evaluation of necessary exceptions that can be approved for a specific period of time. Such exceptions include the use of trade sizes or mechanical connectors that are specified in terms of an inch. Moreover, IEEE implementation of the policy does not require that products specified in terms of an inch be replaced by products specified in metric units.

The discipline of electrical engineering has been extensively metricated for more than a century. In 1881, the British Association for the Advancement of Science (BAAS) introduced practical electrical metric units—the ampere, coulomb, farad, ohm, and volt—to the First International Electrical Congress in Paris (Naughtin, 2008). The ampere was eventually defined as an SI unit of electrical current in 1948 (BIPM, 2006) and was subsequently defined as a base SI unit. The legalities associated with such developments again go to achieving a common understanding of the suggested, optimal, and safe measures affecting how and when to use a technology and of how to correctly perform certain processes based on measurements and values. And, in the case of electrical engineering, even small differences related to interpreting such measurements could lead to danger, disaster, or death in international contexts.

Alliance for the Advancement of Technology (AAT)

Conventional calendar and clock units are only mentioned briefly in the BIPM (2006) SI brochure because the rotational and orbital movements of the Earth are not as invariant as a natural atomic frequency. Nevertheless, conventional calendar and clock units are widely used beyond applications of metrological comparison. Computer algorithms for the formatting of dates in the Gregorian calendar and clock times (per Coordinated Universal Time) are programmed into many computing platforms, including the Unix operating system and the Java programming language.

Methods of formatting dates and times for information interchange are also described in the ISO 8601 standard (ISO, 2004). However, there have been several initiatives to apply decimal or uniform schemes to calendaring or time-keeping. One of these initiatives is the Integrated Chronological Applications System (ICAS) framework of standards for the uniform formatting of calendar and clock information, developed by the Alliance for the Advancement of Technology (AAT). The development of ICAS standards is based on a cross-disciplinary approach that includes technical communication, computer science, usability analysis, applied linguistics, and metrology. Recent ICAS developments have built a framework for determining how calendar and clock units can be formatted or processed in terms of uniform or decimal schemes. Such a framework can provide advantages for users who process or work with a lot of calendar and clock information. And ICAS approaches can inform workflows that decimalize time scales for existing applications in metrology, computer science, astronomy, and information science.

The AAT has also formulated a metrication policy (AAT, 2009d) that tries to highlight a proactive, rather than remedial, and a voluntary, rather than mandatory, approach to promoting metrication. Style and usage guidelines for ICAS are described in the 2020 series of standards documents (AAT, 2009a). Other policies related to the administration of AAT ICAS, such as the terms of use (AAT, 2009c) for the standards documents, or the ICAS in use (AAT, 2009f) resources for communicating ICAS compliance contain many features that are similar to the policies of other standards development organizations and initiatives.

The summary described here is not meant to be a complete list of metrication policies. The policies have however been identified as resources that communicators can use to inform efforts to add value to users of technical information. They also reveal the kinds and the nature of the legal problems that variations in uses of measurement can cause in global contexts. Such potential legal problems, moreover, are complicated by the international diffusion of online media, which allows individuals from around the world to access and use technical documentation regardless of the culture for which those materials were originally designed. Such issues are of particular concern as international online access seems to be growing at a seemingly exponential rate and increasingly exposing individuals to new cultures and new legal systems. For these reasons, readers are encouraged to reference any other policy or initiative that is applicable to a particular situation.

A FOCUS ON BEST PRACTICES FOR METRICATION COMMUNICATION

There are numerous standards and style guides that provide advice that in some cases differs from the advice provided by other sources. As no guide can cover all the style issues for every possible situation, it is important to determine a style-based strategy that includes a library of style resources appropriate to the context in which the guide will be used (Hart, 2000). Particular areas of practice, or particular contexts of use, may also be subject to particular standards. A comprehensive review of existing style and standards resources can, in turn, help an editor to focus on issues that are most pertinent for a particular guide, such as one for the sciences (Hart, 2008). The key in selecting such a guide, in many cases, can come down to how well it accounts for and addresses the prospective legal issues that could be connected to a given situation. As the standards for metric units are revised in light of advances in metrology, there will be a need for current information that has been updated to reflect the state of the art in a way that covers the broadest range of legal concerns related to communicating in global contexts.

Toward developing guidelines for best practices for metrication as related to international technical communication, it may be useful to recognize that a standard or a guideline can serve as an appropriate starting point for guidance, but that it might not have been intended to be a final answer for a particular situation or scope of use. This caveat is essential for interpreting a standard or guideline and can also serve as an important means for developing or improving a standard or guideline. Differences between style guides are not necessarily wrong, however it is also important to be clear about the extent to which and how a particular guideline concerns a matter of scientific or technical standardization or to which it concerns a custom or convention.

In a global context, different writers or readers might bring different approaches, such as regional spelling usage. Standardization in a global setting, however, reflects a certain amount of consensus. In the case of SI, the unit symbols are the same for all languages, and the terms in certain contexts are spelled according to a particular language or region. This is the case for the Special Publication 811 (Taylor & Thompson, 2008) spellings of "meter" and "liter" that differ from the British and French spellings "metre" and "litre." Another style guide, the International English Language Guidelines (INTECOM, 2005) tends to recommend simpler spellings and expressions for a variety of terms. Oftentimes these recommendations reflect U.S. spelling customs; however, for metrication terms, the French and British spellings are strongly advised. BIPM maintains only a French language version of the SI brochure as the official reference, and has also translated the document into English, from which NIST has translated with changes in American spelling for use by the United States (Taylor & Thompson, 2008). For the development of ICAS standards, the Alliance for the Advancement of Technology outlines an approach to accommodating variations in regional spelling: "as a transitional guideline, AAT ICAS elects to accept established regional spellings for purposes of English language localization" (AAT, 2009e).

Another example is the matter of how can one refer to customary or pre-metric measures? The European Community uses the expression "supplementary indications." Should a writer wish to avoid the term *unit* for pre-metric measures such as inch, foot, or mile, then the terms *customs*, *data*, *language*, *measure*, or *practice* can also be used to clearly communicate the status of the usage in terms of a metrication best practice.

What else should be included in a consideration of best practices for metrication communication? Some issues include the language used in software interfaces (is a "kilo" to be defined as 1,000 or as 1,024 when there is an industrial standard for binary prefixes "kibi"), product labeling, and publications. Other issues include needs to be current about matters of metrication and education. Metrication communications should account for best practices and should not be based on outdated generalizations and inaccurate information. Some current theories of audience analysis can be useful in helping writers to select appropriate information for the communication of metrication best practices. Nathan Salmon (1988) describes a "special role" for some measuring words (in this case, the terms *standard*, and *metre* are critically analyzed in the context of some written usage) and highlighting that matters of definition are critical to understanding measure and that some specialized matters of definition might not be commonly understood in the context of everyday usage or out of the context of a particular usage.

Writers preparing guidelines for the communication of metrication best practices should thus strive to promote consideration of best practices for metrication communication over the formulation of rules. As a matter of best practice, the importance, scope, or weight of a guideline should also be clear to users. It may also bear mention that because the particular business situations encountered by communicators can vary considerably, this treatment is not meant to be a complete treatment for any situation. And within this overall framework, technical communicators must always be aware of the web. That is, most forms of documentation will, at some point, be placed online for effective dissemination of information. For this reason, technical communicators need to start thinking of what they do in terms of online delivery and greater global audiences. These two factors—the online and the global—need to be central issues technical communicators use, both when selecting a style guide for and effectively addressing the topic of metrication and the representation of weights and measures in all technical documents. By keeping these factors in mind, technical communicators can select or develop a style guide that can help them more effectively address a range of legal issues in global contexts.

CONCLUSION

Metrication is closely linked to progress in science, technology, and commerce. As science and technology advance, measurements can be realized more accurately. Furthermore, the standardization of a metric system of measures can better accommodate best practices of measure. In commerce, the metrication benefits for producers and consumers can offer more value than would be possible with pre-metric customs of measure. Perhaps the methods by which uses of metric units can simplify matters of "great difficulty" in addressing problems of measurement is why frameworks of metrication policy have been widely established by governments, researchers, and industries. This viewpoint also takes into consideration both proactive and any remedial costs associated with a particular program of metrication. However, once metrication is attained, there are no remedial costs to hinder productivity or quality.

Yet metrication does call for education and training about the use of metric units of measure, and the quantity of information on metrication can be overwhelming. There are several methods by which technical communicators can deliver "best practices" for metrication communication. Rhetorical analyses such as determining the audience, purpose, and context of use for a particular document can guide a communicator to develop and produce information that is more useful (Hart, 2000). A document prepared for a particular area of practice or for a particular local audience may call for some approaches that provide additional or specific guidance. At the same time, metrication and science are practiced globally among several areas of practice. If metrication can add value to the producers and the users of communications, then there are incentives and opportunities (and in some cases also requirements) to provide metrication communication.

REFERENCES

AAT. (2009a). AAT ICAS 2020: Usage guidelines. *Alliance for the Advancement of Technology*. Retrieved from http://www.aatideas.org/icas/2020.xht

AAT. (2009b). AAT ICAS 2040: Guidelines for scale calculation. *Alliance for the Advancement of Technology*. Retrieved from http://www.aatideas.org/icas/2040.xht

AAT. (2009c). AAT ICAS 9010: Terms of use. *Alliance for the Advancement of Technology*. Retrieved from http://www.aatideas.org/icas/9010.xht

AAT. (2009d). AAT ICAS 9030: Metrication policy. *Alliance for the Advancement of Technology*. Retrieved from http://www.aatideas.org/icas/9030.xht

AAT. (2009e). AAT ICAS 3005: Note on language support. *Alliance for the Advancement of Technology*. Retrieved from http://www.aatideas.org/icas/3005.xht

AAT. (2009f). About ICAS in use. *Alliance for the Advancement of Technology*. Retrieved from http://www.aatideas.org/iota/icas/icas.xht

Aristotle. (1991). *On rhetoric: A theory of civic discourse* (G. A. Kennedy, Trans.). New York, NY: Oxford University Press.

Becker, P., De Bièvre, P., Fujii, K., Glaeser, M., Inglis, B., Luebbig, H., et al. (2007). Considerations on future redefinitions of the kilogram, the mole and of other units. *Metrologia, 44*(1), 1–14.

BIPM. (2006). *The international system of Units (SI)* (8th ed.). Sèvres Cedex, France: Bureau International des Poids et Mesures.

EEC. (2000, February 9). Council Directive 80/181/EEC. CONSLEG: 1980L0181. *Office for Official Publications of the European Communities.* Retrieved from http://eur-lex.europa.eu/LexUriServ/LexUriServ.do?uri=CELEX:31980L0181:EN:HTML

EU. (2007). Explanatory memorandum on proposal for a directive of the European parliament and of the council amending Council Directive 80/181/EEC on the approximation of the laws of the member states relating to units of measurement. *European Union.* Retrieved from http://eur-lex.europa.eu/LexUriServ/LexUriServ.do?uri=CELEX:52007PC0510:EN:HTML

Gómez-Lobo, A. (1978). Aristotle's first philosophy and the principles of particular disciplines: An interpretation of metaphysics E, 1, 1025 b 10–18. *Zeitschrift Für Philosphische Forschung, 32*(2), 183–194.

Guinot, B. (1997). Application of general relativity to metrology. *Metrologia, 34*(3), 261–290.

Hart, G. (2000). The style guide is dead: Long live the dynamic style guide. *Intercom, 47*(3), 12–17.

Hart, G. (2008). Scientific style and format: The CSE manual for authors, editors, and publishers. *The Exchange, 15*(2), 9–11.

IEEE. (2007). *IEEE standards style manual.* New York, NY: Institute of Electronics and Electrical Engineers.

ISO. (2004). *ISO 8601:2004: Data elements and interchange formats—information interchange—representation of dates and times* (3rd ed.). Geneva, Switzerland: International Organization for Standardization.

Juengel, M. (2011). Wrongful death from medication overdose—Math ability exams screen competencies. *Prophecy.com.* Retrieved from http://www.prophecyhealth.com/wrongful-death-from-medication-overdose-math-ability-exams-screen-competencies/

Language requirements in Europe. (2012). *Net-Translators.* Retrieved from http://www.net-translators.com/knowledge-center/articles/language-requirements-in-europe

Mills, I. M., Mohr, P. J., Quinn, T. J., Taylor, B. N., & Williams, E. R. (2006). Redefinition of the kilogram, ampere, kelvin and mole: A proposed approach to implementing CIPM recommendation 1 (CI-2005). *Metrologia, 43*(3), 227–246.

Naughtin, P. (2007). Commentary on John Wilkins' of measure. *Metricationmatters.com.* Retrieved from http://www.metricationmatters.com/docs/CommentaryOnWilkinsOfMeasure.pdf

Naughtin, P. (2008). Metrication timeline: A chronological history of the modern metric system. *Metricationmatters.com.* Retrieved from http://www.Metricationmatters.com/docs/MetricationTimeline.pdf

NIST. (1998). Metric system of measurement: Interpretation of the international system of units for the United States. *United States Federal Register, 63*(144), 40334–40340.

Reid, J. B. (2008a). Metrication of Australia. *United States Metric Association.* Retrieved from http://lamar.colostate.edu/~hillger/internat.htm#australia

Reid, J. B. (2008b). Canadian metrication. *United States Metric Association.* Retrieved from http://lamar.colostate.edu/~hillger/internat.htm#canada

Salmon, N. (1988). *How to measure the standard metre.* Proceedings of the Aristotelian Society. Vol. LXXXVIII (pp. 193–217). Oxford, UK: Aristotelian Society.

Stephenson, A. G., LaPiana, L. S., Mulville, D. R., Rutledge, P. J., Bauer, F. H., Folta, D., et al. (1999, November 10). Mars climate orbiter mishap investigation board: Phase I report. *NASA*. Retrieved from ftp://ftp.hq.nasa.gov/pub/pao/reports/1999/MCO_report.pdf

Taylor, B. N., & Thompson, A. (2008). Guide for the use of the international system of units (SI). Gaithersburg, MD: Institute of Standards and Technology.

Tondering, C. (2008). Frequently asked questions about calendars. *The Calendar FAQ*. Retrieved from http://www.tondering.dk/claus/calendar.html

U.S. Department of Health and Human Services. (2001). *Cultural competence works: Using cultural competence to improve the quality of health care for diverse populations and add value to managed care arrangements.* Washington, DC: Health Resources and Services Administration, DHHS.

U.S. Metric Association. (2005). *Metric convention of 1875.* Retrieved from http://lamar.colostate.edu/~hillger/laws/metric-convention.html

Wignall, J. W. G. (2007). Some comments on the definition of mass. *Metrologia, 44*(3), L19–L22.

PART II

Language and Access

http://dx.doi.org/10.2190/LIIC5

CHAPTER 5

Legal Literacy for Multilingual Technical Communication Projects

Tatiana Batova

There are many incentives for technical communicators to think about legal issues in multilingual projects. Because many importing countries (e.g., countries of the European Union and Russia) require user documentation to be translated into their official languages, and there is a significant growth in U.S. exports,[1] the volume and costs of technical translations are rising. At the same time, since consumer protection and translation laws are becoming better defined and more stringent (e.g., EU Resolution C411), substandard translations are increasingly likely to cause liability danger and become a source of litigation. Even more importantly, international businesses have a legal and ethical duty to make their products safe for the consumers. These businesses need to devote resources to making product information adequate for international consumers. They must do so to protect themselves from liability litigation, to encourage import/export, and to meet "ethical standards of utility, rights, justice, and care" (Markel, 2001, as cited in Lipus, 2006, p. 76).

In this chapter, I look at legal issues in multilingual technical communication projects. I argue that the concept of legal literacy that emphasizes the civic engagement of technical communicators and critical rethinking of legal contexts is very useful for discussing the legal issues of international technical communication and needs to include these issues. I start by reviewing how the focus of discussions of legal issues in technical communication has been changing over the last 30 years. I then point out that two of the newest directions—seeing technical communicators as co-producers of law and including global contexts

[1] According to St. Louis Federal Reserve data service (FRED, 2013), there was a 91% growth in U.S. exports between 2000 and 2011.

into the discussion of legal issues—have a great deal in common. I then use an adapted framework of two core acts of legal literacy in technical communication outlined by Hannah (2011) to summarize issues of international technical communication. In so doing, I discuss the contributions that technical communicators, translators, and law authors can make to the idea of legal literacy in global contexts. Through this discussion, I elucidate what legal literacy for multilingual technical communication projects entails, and I outline possible directions for future research.

THE CHANGING FOCUS OF LEGAL QUESTIONS IN TECHNICAL COMMUNICATION

The relationship between technical communication and law has attracted consistent interest in our field. For the last 30 years, the focus of this interest has varied and has included the following areas:

- Examining product liability concerns (English, 1995; Helyar, 1992; Smith & Shirk, 1996)
- Complying with standards and government regulations in the United States (Collins, 1999; Strate & Swerdlow, 1987)
- Interpreting the definitions of commonly used legal terms and phrases in product liability cases (Bedford & Stearns, 1987; Ross, 1981)
- Avoiding defective instructions (McGuire, 1988)
- Making instructions noticeable (Wogalter, Barlow, & Murphy, 1995)
- Creating effective safety instructions, safety labels, and warnings (Madden, 1999; Ross, 1981; Smith, 1990; Velotta, 1987)
- Understanding implications of legal writing (Goodwin, 1996; Griggs, 1996; Moxley, 1996; Taylor, 1996)
- Considering ethics (Bailey, 2007; Markel, 2001; Porter, 1987) and ever-changing technology (Herrington, 2003) in the discussion of legal issues in technical communication
- Exploring the role of intellectual property in scientific/technological development (Bazerman, 2002; Durack, 2006; Reyman, 2010)
- Applying rhetorical theory to the study of the law (Ranney, 2005)
- Researching law's influence on the construction of power (Britt, 2001; Lay, 2000; Propen & Schuster, 2008; Schuster & Propen, 2010)

A new wave of interest in the topic was apparent in two special issues of major technical communication journals. In 2010, the journal *Technical Communication* (*TC*) published its special issue "Legal Issues in Global Contexts: Reconsidering Content in an Age of Globalization," and in 2011, the journal *Technical Communication Quarterly* (*TCQ*) followed with its "Technical Communication and the Law." These two special issues led technical communicators in two interesting directions.

Contributors to the *TCQ* special issue emphasized the importance of civic engagement within technical communication and the need for technical communicators to start thinking of themselves not just as consumers but also as co-producers of law. To help technical communicators see the legal potential in the rhetorical situations of composing and to free them from confining their roles to victims of legal limits and restrictions, Hannah (2011) introduced the concept of *legal literacy*. Legal literacy encourages technical communicators to view their work as "imbricated" in law and to view law as "imbricated" in the world. For Hannah, this legal literacy consists of two core acts: seeing the law in technical communication and understanding the law in technical communication. This dual framework, I argue, calls for development of competency in the following three areas:

- Area 1: Combining a general awareness of legal issues with knowledge of the context in which law is created. This includes skills related to navigating law and applying it to composing practices, as well as developing procedural understanding through legal reasoning, legal writing, and legal research.
- Area 2: Recognizing opportunities for co-producing the law by setting precedents and "resisting or pushing against regulations, guidelines, or laws" (Hannah, 2011, p. 13) while composing.
- Area 3: Learning to engage in predictive thinking and writing and to uncover "legal potentiality" (Hannah, 2011, p. 12)—to analyze unintended and unforeseen interactions with a product or documentation by the user.

Spilka (2010) points out that technical communication authors (Cargile-Cook, 2002; Kastman-Breuch, 2002; Nagelhout, 1999; Selber, 2004; Slattery, 2007) increasingly propose the existence of multiple literacies or multiple layers of literacy in work contexts. Cargile-Cook (2002), for instance, argues that "today, technical communicators need to be multiliterate, possessing a variety of literacies that encompass the multiple ways people use language in producing information, solving problems, and critiquing practice" (pp. 5–6). According to Hannah (2011), central to Cargile-Cook's claim is the awareness that technical communicators work at an intersection of many competing interests. In multilingual projects, the work of technical communicators overlaps with the work of a number of parties, including freelance translation/localization specialists, translation agencies, legal departments of their company within the country and abroad, and regulatory authorities and standardization organizations. The intersection of the interests of these groups is often complicated by the multiple legal concerns of the globalized market into which products are launched.

The authors whose articles appear in the special issue of the *TC*, by contrast, focus on the legal concerns that surround content creation and content management in the globalized market. As the editors of the special issue note in their introduction, both content creation and content management processes have certain frictions. This is because they are influenced by human constructs that reflect deeply rooted perspectives of what behavior is considered appropriate by different groups. These perspectives stem from culture, and they are often codified into laws that regulate the production and the sharing of information (St.Amant & Rife, 2010). The editors of the special issue also argue that developing practices to address friction points in international interactions is a way for technical communicators to add value to their organizations and can thus serve as a strategy for achieving job security and for opening opportunities for advancement.

Articles in both *TCQ* and *TC* special issues focus on developing certain competencies in relation to law and communication. When viewed from a global legal context, translation emerges as a key area related to such competencies. That is, with international markets growing, technical communicators are increasingly involved in projects that require translation. Even though technical communicators today are urged to develop new competencies continuously, getting more insight into the translation industry and its legal issues is becoming a must. And this emphasis only grows as more organizations focus their business strategies on international markets.

LEGAL ISSUES IN MULTILINGUAL TECHNICAL PROJECTS

With an emerging, robust body of quality technical communication research on legal issues, we now need to systemize the legal concerns in the field in order to make the results of this research more readily suitable for instruction and practice. In this section, I argue that the concept of legal literacy can help encompass the levels of knowledge and understanding needed for technical communicators to make legal issues more manageable. To begin this process, one first needs to define the concept of *multilingual legal literacy*. For the purposes of this chapter, I use an adaptation of the two core critical acts of legal literacy presented by Hannah (2011). For international technical communication, these two core acts result in the following three competency areas:

- Navigating and understanding the related laws and the contexts in which they are applied and interpreted,
- Resisting/co-producing the law, and
- Uncovering the legal potentiality of technical translation.

Laws and Standards: Navigating and Understanding the Legal Field

Intercountry business operations are controlled by a combination of international regulations and multiple country-specific laws and standards that can vary between nations. The differences in national legal systems and nation-specific laws and standards can be daunting. As a result, they are often left to legal departments to tackle. However, as Kelly and Jones (2007) argue, legal issues are also localization issues.[2] They must therefore factor into multilingual projects "not just during the legal review at the end, but during the localization process, just as design, layout, graphics, and rhetoric are" (p. 204). To include legal issues in the design of multilingual texts, technical communicators need to know how to navigate the intricate interplay of various national regulations that govern product liability and the necessity of translation. Similarly, they need to understand the different standards that prescribe rules for achieving quality translation.

Necessity of Translation

The necessity of translating technical documents in the United States varies for different products and services. In the healthcare fields, for example, there is a legal requirement to provide patients who have a limited understanding of English with translation and interpretation services (St. Germaine-McDaniel, 2010). Under Executive Order 13166, persons who have limited English proficiency (LEP) should be provided with meaningful access to healthcare services including any necessary written information. This factor makes technical communicators a key professional group involved in making "health care information accessible to diverse populations" (St. Germaine-McDaniel, 2010, p. 252).

It is true that not every sphere of services and goods in the United States is regulated as strictly as health care. Several U.S. legal cases, however, put the necessity to provide product information in languages other than English into perspective. In one 1992 legal case (*Stanley Industries Inc. v. W.M. Barr & Co. Inc.*), a factory owner in Miami, Florida, sued the manufacturer of linseed oil after a fire was caused by a spontaneous combustion of rags saturated with the oil. Even though the English-language label on the product warned against such a possibility, one of the employees of the factory spoke only Spanish, and another had only limited English proficiency.

The plaintiff in this case argued that because the warning did not address Spanish speakers, the product was defective. In so doing, he presented evidence that the manufacturer advertised the product to Spanish-speaking audience by, for example, placing an ad for linseed oil in a Spanish-language newspaper (even though the ad was to begin three days after the fire). In this case, the court ruled that because the defendant advertised in the Hispanic media, "genuine issues of fact existed regarding the warning's adequacy" (Vignali, 1995, p. 85). Moreover, the jury was instructed to decide if the manufacturer could have

[2] Localization involves "taking information that is technically and culturally specific to the original language and changing it into a form that is correct and meaningful to other countries" (Hackos, 1994, p. 497).

foreseen the use of the product by non-English speaking people and determined if the English-language warning was adequate.[3] In addition to the idea that the absence of a Spanish warning could cause serious harm and damage, the manufacturer carried a serious litigation burden in this case. As this example illustrates, even when a product is to be sold in the United States, and even if there might be no specific regulations prescribing translation, it is important to think critically about translating user documentation.

When products are exported, the necessity of translation often becomes more regulated. European law, for example, requires all product documentation to be translated into the language of the countries in which it is sold (EU Resolution C411, 1998). According to this resolution, technical products can be complete and ready for distribution only if they are accompanied by operating manuals in the respective national languages. Moreover, as Byrne (2007a) notes, liability due to incomplete documentation is incorporated into the laws of most EU member states. Germany's product liability law (Produkthaftungsgesetz), for example, entitles users to replacements and/or compensation if the documentation results in damage or loss (Heino, 1992, as cited in Byrne, 2007a).

Adequate Translation Quality

Not only does the EU Resolution C411 (1998) require translation, it also aims to improve the quality of technical translations by stating that the documentation must be clear, accurate, and easily comprehensible. It also explains, to a certain degree, what "clear, accurate and easily comprehensible" means (i.e., such documentation must be designed for a specific audience to help this audience use a product safely and correctly). Thus, translated operating instructions that do not provide adequate help in performing a task can result in the determination that a product is defective. Such findings, in turn, can be a source of liability and a foundation for claims for compensation (Byrne, 2007a).[4]

Within such contexts, translation-related standards are very helpful for achieving translations that are "easily understood and applied by the user" (98/79/EC, in-vitro diagnostic medical devices). For the EU, mechanisms that can help

[3] For a more detailed description of this case and additional examples of similar cases, see Vignali, 1995.

[4] Byrne's article (2007a, available at http://www.jostrans.org/issue07/art_byrne.php) and his book "Technical Translation: Usability Strategies for Translating Technical Documentation" (2010) can be a good source to research laws of the EU that deal with translation requirements for different product areas (90/385/EEC and 93/42/EEC for medical devices, 76/768/EEC for cosmetic products, 88/378/EEC for toys, 90/396/EEC for appliances burning gaseous fuels). Hoft's *International Technical Communication: How to Export Information about High Technology* is also an invaluable resource for researching standards (we need to be careful, however, since many changes have happened in the legal field since the publication of the book in 1995).

achieve such translation-related objectives include selected ISO, ASTM, DIN, EUATC, UNI, ÖNORM standards.[5] They can also include quality metrics established by the SAE (J2450) and the quality assurance (QA) model used by the Localization Industry Standards Association (LISA). When used as part of translation contracts between the clients (technical communicators) and translation services providers (agencies or freelancers), these standards can be a valuable resource for achieving translation quality.[6]

Generally, the conversation about translation contracts is common in the translation literature. These works, of course, look at translation contracts from a translator's perspective. They concentrate on things such as determining payment schedules and limiting a translator's financial liability for a company's losses that result from a substandard translation. For example, Ansaldi (1999) recommends that "translators include in their terms of business a clause that explicitly excludes liability for consequential losses and entitles the client to a refund alone" (Byrne, 2007a). To work effectively with translators, technical communicators must therefore understand translation contracts and be able to ground their composing decisions in their knowledge of the translation industry and its complexities.

Product Liability

Lipus (2006) similarly argues that, in addition to understanding translation laws and standards and being able to use this understanding when creating translation contracts, "technical communicators who write product documentation for users in different countries must understand the product liability and consumer protection issues of those countries" (p. 77). She points out that technical documentation should be viewed as a safety link between exported goods and their international consumers. She also offers an excellent review of the consumer protection measures that major U.S. trade partners have implemented. (These partners include Canada, the EU, Mexico, Japan, China, Korea, Taiwan, Singapore, Hong Kong, Australia). Among the legal factors that

[5] ISO 2384: Documentation. Presentation of translations; ISO 704: Terminology work. Principles and methods; ASTM F15.48: Consumer-oriented guide to quality assurance in translation and localization; DIN 2345: Translation contracts; ÖNORM D 1201: Translation and interpretation services. Translation services. Translation contracts; EUATC: Quality standard for translation services.

[6] An excellent resource to start getting acquainted with these standards is Uwe Muegge's book, *Translation Contract. A Standards-Based Model Solution.* Muegge's goals are to improve "the communication between translation vendors and translation buyers," structure and standardize translation projects, and improve the efficiency and the quality of translation projects. The book contains multiple components of an ideal translation contract and is aimed at "translation buyers and vendors who do not have much experience in the translation and/or localization field" (2005, p. v).

technical communicators need to consider when developing materials for international audience, Hoft (1995) lists trademarks and copyrights; licenses and warranties; health, personal safety, and environmental requirements; and labels on machinery.

While the articles and books mentioned previously are helpful for researching laws and standards pertaining to technical translation, another good resource can be providers of translation services, as well as employees of legal departments. These international technical communication stakeholders can direct technical communicators to specific resources and can help them become more acutely aware of differences in legal systems (e.g., civil versus common law). Such differences imply that varying legal requirements pertain to technical documentation, and that global audiences are situated in different legal contexts. Technical communicators must remember that particular legal contexts will result in certain reader expectations for technical texts, especially those related to culture. When these expectations are not met, reader response to technical texts can be less favorable than technical communicators might want.

Culture and User-Centeredness: Resisting and Co-Producing Law

When documentation for products is distributed internationally, laws and regulations become inseparable from the cultural variables and the expectations of international audiences (Lipus, 2006). Complying with laws and following the standards is often not enough to reach a desired audience response, since "difference in culture may cause the documents to be interpreted the way the US supplier did not intend" (Ulijn, 1996, p. 70). In fact, several researchers in technical communication have noticed and discussed situations in which legal and cultural expectations of the readers contradict legal and cultural expectations of the writers. These authors point out challenges that such situations impose, as well as strategies for overcoming them and achieving a user-centered design.

In their work on legal issues in website localization, Kelly and Jones (2007) state that people can understand things differently based on their intrinsic legal culture. These authors single out disclaimers on websites as one area of potential problems, particularly when comparing responses to translated websites between users from countries with case law and those from countries with civil law. Kelly and Jones come to the conclusion that consumers in common law countries (e.g., the United States) take disclaimers for granted and do not view them as likely or immediate outcomes of an action. However, in civil law countries (e.g., France), where consumers are not so inundated with disclaimers, they can view disclaimers as statements of probable outcomes. Thus, an uncritical use of disclaimers for readers acculturated in a civil law system can make a website culturally inappropriate and can damage the perception of a product's reliability (without offering any additional legal protection).

As I describe elsewhere (Batova, 2010), some legal presuppositions are often taken for granted in a monocultural context and could present unexpected challenges in an international environment. For example, in an international clinical trial setting, patients' expectations for family participation in medical decision-making could clash with regulations governing the privacy of medical information in the country where a particular clinical trial originated. In such cases, uncritical compliance with law disregards "cultural differences in decision making" (Kim, Young, Neimeyer, Baker, & Barfield, 2008, p. 352) and the needs of the audience.

Pander Maat and Lentz (2011), critical of regulations when thinking of user-centered design, reviewed the patient information leaflets and their compliance with a template published by the Quality Review of Documents (QRD) group of the European Medicines Agency (EMEA). Pander Maat and Lentz found that one leaflet, which is available in 25 European languages (the English version has precise instructions for what type of information should be placed under the various headings), does not match user expectations. In fact, there were two mismatches: the first mismatch was between the wording of headings and reader interpretations; the second mismatch had to do with classifying and grouping information. Such mismatches make the findability of information poor, and they can contribute to decreased reading performance.

While the three works discussed above call on technical communicators to be critical of laws and regulations when creating user information, McKee and Porter (2010) look at technical communicators as consumers of information. Reviewing the ethical and regulatory issues that they believe technical communicators need to reflect on when conducting Internet research, McKee and Porter emphasize the differences in legal regimes and privacy laws and the complexity of intellectual property on a global scale. They show that technical communicators need to understand "the cultural expectations of the persons whose communications they are studying" (p. 296) and remain flexible in their research design.

The topics of international law may differ for the authors in this section, but a key theme in their work is that technical communicators need to

- Be aware of contradictions,
- Be critical of international and county-specific regulations,
- Keep the interests of international users in mind, and
- Strive to influence change.

These tasks are not easy to achieve, but they do represent an ideal that technical communicators can strive to accomplish.

Creativity, Liability, and the Task of the Translator: Uncovering Legal Potentiality

Most of the new tasks noted here are hard to achieve without input from another work group that plays a vital part in multilingual technical projects—technical translators. Not only do translators speak the languages of international users, they also share users' cultural and legal contexts. In ideal situations, technical translators are cultural consultants and a great informational resource. In less-than-ideal situations, they are viewed as contributors to translations that are neither user-friendly nor accurate. To avoid such negative perceptions, to avoid situations with legal implications, and to focus instead on strategies for successful collaboration, technical communicators need to understand the legal issues of the technical translation industry.

Legal potentiality in technical translation stems from a set of innate contradictions that characterize translation as a field of study and technical translation as an industry. On the one hand, technical translation is a creative process; on the other hand it is a service. Translation work leaves many decisions to each individual translator; however, technical translation needs to be strictly regulated. To understand how these contradictions create legal potentiality in multilingual technical projects, we need to look at how the creativity of technical translators is connected to their liability. We must also examine how even a general understanding of the tasks of technical translators will result in a recognition of complexities with translation evaluation.

Creativity and Liability

Translation has long been considered a "heuristic process that involves discourse analysis and interpretation, analogical thinking, and re-formulation and re-expression of the ideas of the source text" (Delisle, 1980, as cited in Weiss, 1995, p. 411). It is a complex thought process that in its "expansive and creative way . . . contributes to the growth of its own language" (Munday, 2008, p. 169). While technical translation has originally been seen as involving less original work, several authors have argued that it is a creative practice as well (Byrne, 2009; Pommer, 2008), since the whole concept of translation "calls into question claims of singularity of meaning (absence of ambiguity) in all texts, including business and technical communication texts" (A. Benjamin, 1989, as cited in Weiss, 1997, p. 327). Experiments with machine translation have also shown that technical translation is a very sophisticated process that requires more than transferring the meaning of each word and the knowledge of technical terminology.

Viewing technical translation as a creative process automatically grants technical translators more freedom with the product of translation—a translated technical text. At the same time, becoming creative producers of information means more recognition, and recognition means more liability (Hammond,

1995). If translation is viewed as a creative process that "involves some form of intellectual addition to or processing of the information in the source text, whether by adding, removing, clarifying, interpreting, rephrasing, recontextualizing or recasting information for the target audience," technical translators become subject to a great degree of liability for the information they produce (Byrne, 2007a).

Even though technical translation is rapidly gaining the status of a practice that involves creativity and judgment (Ansaldi, 1999), technical translators are often viewed as service providers. Technical translators are then expected to work in a "good and workmanlike manner" (Ansaldi, 1999, as cited in Byrne, 2007a): "in accordance with the skills, practices and customs appropriate to the profession generally and appropriate to the ordinary skills and capacity normally possessed by a practitioner" (Mowat, 1998, as cited in Byrne, 2007a). While this definition provides some grounds for determining the liability of a translator, Mayer-Schöneberger (1999) argues that the work of translators is then evaluated not by the correct outcomes of their work but by whether translators used their best efforts while translating. As a result, provided a translator can prove that the translation was produced in an appropriate and careful manner to the best of the translator's ability (Ansaldi, 1999), errors may be viewed more leniently and the degree of a translator's liability becomes quite obscure.

With a rather muddled definition of a translator's liability, there is little evidence to date of court cases involving translations of technical documentation (Ansaldi, 1999; Byrne, 2009; Chriss, 2006; Hammond, 1995). At the same time, according to SDL International,[7] "eight out of 10 international businesses are suffering because of translation errors, causing lost revenue, delayed product launches or even fines for non-compliance" (Translation errors, 2007). Why is there such a disconnect? It might not be financially viable to sue a stand-alone freelance translator. Similarly, translation agencies might include clauses in their contracts that limit their financial liability, and many translation disputes might be settled out of court. However, to determine if a translation is incorrect, one needs to define the original task of the translator.

The Task of the (Technical) Translator and Translation Evaluation

"The task of the translator," the eponymous concept from the founding work by Walter Benjamin (2008), has undergone significant change over time, yet is still widely discussed in translation studies. The task of the translator traditionally relates to defining how true to the original text a translator should

[7] SDL International is one of the largest producers of computer-assisted translation systems; now they produce content management systems as well.

be and how much change to the text a translator can make. Originally, Walter Benjamin wrote that

> the task of the translator consists in finding that intended effect [*Intention*] upon the language into which he is translating which produces in it the echo of the original. . . . A real translation is transparent, it does not cover the original, does not black its light, but allows the pure language, as though reinforced by its own medium to shine upon the original all the more fully. (pp. 19–20)

Thus, the role of a translator is to always follow the source language and bring its elements into the target language.

However, conflicting views led many translation authors in a different direction. According to Venuti (1995), translators strive to create the appearance that a translation is not a translation, but the "original," because a "translated text, whether prose or poetry, fiction or non-fiction, is judged acceptable by most publishers, reviewers and readers when it reads fluently, when the absence of any linguistic or stylistic peculiarities makes it seem transparent" (p. 1). Later approaches called translators to think about audience's needs and expectations (relevance theories), as well as the purpose of the target text together with the situational and cultural background of the audience (e.g., Skopos theories). Such perspectives lead many translation practitioners toward localization, a process of creating a text that is linguistically and culturally appropriate for the target language audience.

Unfortunately, these translation theories, which are a basis for translator education, provide little help for technical translation practitioners. For literary translation, these theories result in differing translations of a single literary work that could be masterpieces in their own right. Technical translators are left with little practical guidance to make their own decisions, "which can be regarded as, at best, reasoned but ad hoc or, at worst, subjective" (Byrne, 2010, p. 45). The work of technical translators, however, needs to be regulated and judged, because "readers of a translated legal contract, an informed consent, or a user's manual" have "quite different requirements, as these documents can directly affect their well being or their ability to use a particular product" (Stejskal, 2006, p. 12). Hammond (1995) cited a case from Switzerland (described by Holz-Mänttäri in 1987) that provides us with an interesting insight into the role of a technical translator and its legal implications. In this case, a client did not pay a translator because the translation of a technical document "did not resemble the source text closely enough" (p. 233). The translator, in turn, sued the customer. The court determined that a translation is good if it does the job it was prescribed to do and ordered the customer to pay the translator. In essence, because the source text was not appropriate for its intended purpose (product specification), the translator had to base her translation not on the source text provided by the customer but on the appropriate product specifications (which was exactly what she did).

Striving to outline more tangible ways to regulate the work of technical translators, the American Translators Association (ATA, n.d.) published a brief translation buying guide in which the quality of a translation is equated to "the degree to which it follows the agreed-upon specifications" (p. 4). The authors of the guide advise using translation industry standards (ASTM, ISO, EN, CEN, DIN) to develop these specifications. They also point out that the client (technical communicators) should agree with a translation service provider (translation agency) on a set of specifications. These contracts could be a valuable resource for translation projects, but they could also lead to evaluating translation not by the final product, but rather by the process of translation.

One part of the task of the translator as a professional practitioner is a duty of care to the client (technical communicator). Byrne (2007a) refers to the *Caparo Industries v. Dickman* case (1990) to provide the definition of duty of care. This duty of care presupposes the following components:

- "The information is provided for a purpose."
- "This purpose is made known at the time of commissioning."
- "The information provider knows that the information will be provided to the recipient in order to be used for the specified purpose."
- "It is known that the information is likely to be used without independent inquiry or verification for the stated purpose."

Such an understanding of duty of care means that translators should always know the purpose of the text. Even though professional translators should ask the client for this purpose, if their client cannot provide such information, there is nothing a translator can do. In addition, this understanding of duty of care implies that the translators' duty of care is diminished if a translation they provided is used for a purpose different from the originally specified (adapted from Byrne, 2007a).

Not surprisingly, with so much contradiction around the role of a translator in translation, the view of a translators' task has been shifting in technical communication. Hallman (1990), for instance, notes that "a technical translator's only recourse . . . is to provide a faithful rendering of the text" (p. 245). Even if the translation is bad but stays true to the source (original) text, the translator's work is accomplished. Eubanks (1998) stresses that while translators do make composing decisions, they are "primarily concerned with fidelity to a source text" (p. 52). Maylath (2001) adds the usability variable to describe translation quality. The author, however, still uses the term *accurate* in this context (while *accurate* means free from error, it also has the connotation of *conforming*). Weiss (1995) cites Sándor and Higgins to note that technical translation is pragmatic, practical, and purposive, and thus focuses on the target language and audience the most as compared with other types of translations.

For technical communicators, the ambiguous understanding of the task of the translator in translation evaluation raises many questions with legal implications. If the original text is bad, does it mean that the translation also has to be? If a translation deviates from the original document, does that mean it's bad? Can we agree that a translator performed a task well if a translation does not meet the cultural requirements of the readers? Who would be responsible for making the call for localization? How can we determine the responsibility for a translation that doesn't comply with the law or is not user-friendly for a specific culture? What happens to liability when texts are repurposed? While these questions do not yet have a definitive answer, technical communicators need to take a more active role in answering them for specific translation projects (translators want to succeed and establish a long-term working relationship, but without critical information from technical communicators this success would be next to impossible).

LEGAL LITERACY FOR MULTILINGUAL PROJECTS: TECHNICAL COMMUNICATORS AS ACTIVE PARTICIPANTS IN LEGAL DISCOURSE

Being able to formulate these questions and answer them, even to some degree, is a move toward developing legal literacy for multilingual technical communication projects. In a multilingual project, the outcomes of translation are often not considered to be the responsibility of technical communicators. When problems occur, the all-too-common question is then, "Who is responsible and whom should we blame for unsuccessful translated technical documents?" Legal literacy provides a different route. That is, we could try to find out instead how and why a translated document was unsuccessful and try to determine if we could have foreseen this situation to begin with. In other words, we need to think of the legal literacy for multilingual technical communication projects in terms of problem-solving competencies. Because the competencies in the three areas of multilingual legal literacy encompass a wide range of cultural variables, regulations, and work groups, mastering them is not a simple matter. In Table 1, I outline specific tasks that technical communicators will need to accomplish to develop each competency.

The type of legal literacy described in Table 1 is useful for both technical communicators and technical translators. It allows technical communicators to see the importance of their role in influencing the success of translation projects. It also allows technical translators to see how they can be introduced to legal discussion in technical communication. Both groups are able to see how their efforts are interconnected within a legal context.

Table 1. Legal Literacy for Multilingual Projects: Competencies and Tasks

Competencies	Tasks
A general awareness of legal issues in multilingual technical communication projects and the skills for navigating laws and standards.	Learning how to navigate international laws and whom to ask for help with international laws. Remembering that laws/standards may differ; never making assumptions without research.
Recognition of the possibility that contradictions exist between law and culture or the principles of user-centered design; the ability to use this recognition as an opportunity for co-producing the law.	Becoming critical of laws when they disregard the needs of the readers. Developing the knowledge base for arguing the need to change contradicting regulations. Developing strategies for influencing change.
The ability to uncover the legal potentiality in translation and using this ability as grounds for discussion with translation providers.	Starting to think and write "in a predictive manner" (Hannah, 2011, p. 16). Focusing on avoiding substandard translations: providing translators with translation contracts with specific project details, including the purpose and the audience of the translation. Knowing how to use translation industry standards to compose translation contracts; developing knowledge base for understanding/altering translation contract templates; discussing QA procedures of the translation agency and knowing what to ask for. Using translation contracts not just as liability protection, but as grounds for discussion and negotiation. Understanding the complexities of the translation and their legal ramifications (e.g., picking a qualified translator, translation evaluation) and using this understanding to evaluate legal potentiality of translation projects. Viewing translators as colleagues who have the same goal and not as people who might cause legal problems; knowing that this includes thinking of translators as cultural consultants and including them in projects early on.

FURTHER DIRECTIONS FOR DEVELOPING LEGAL LITERACY FOR MULTILINGUAL TECHNICAL COMMUNICATION PROJECTS

This chapter examined legal complexities related to multilingual technical projects. The chapter also defined the concept of multilingual legal literacy. Work on multilingual legal literacy, however, should not stop here. To continue exploring these ideas, technical communication researchers should consider

- *Including technology into the conversation of multilingual legal literacy.* What are the implications and legal consequences of content management for translation projects? Content management allows repurposing text, and technical translators need to know the exact purpose of the text to guarantee translation quality.
- *Refining strategies for obtaining information on legal differences between countries and the influence of these differences on the relation of reader to text.* Because one multilingual project can include translation into several languages, it can be impractical to require a technical communicator to research each country.
- *Making results of research on legal literacy for multilingual technical communication projects more readily available to practitioners.* Researching legal issues and topics can take months; that is why practitioner-oriented materials need to be created and disseminated.

Developing the concept of multilingual legal literacy also requires more communication between technical communicators and technical translators. Equally important is the realization that we need to pay even more attention to technical translation in our field, because "the legal importance of a good translation and the savings gained from it indicate that its costs are a necessary and worthwhile investment which will be repaid at a later stage" (Révy Von Belvárd, 1997, p. 191).

REFERENCES

American Translators Association (ATA). (n.d.). *Translation: Buying a non-commodity.* Retrieved from http://www.atanet.org/docs/translation_buying_guide.pdf

Ansaldi, M. (1999) Translation and the law: Observations of a law professor/translator. *Language International, 11,* 12–17.

Bailey, N. J. (2007). Why you should be worried about more than just a lawsuit. *IEEE Engineering Management Review, 35,* 104–105.

Batova, T. (2010). Writing for the participants of international clinical trials: Law, ethics, and culture. *Technical Communication, 57,* 266–281.

Bedford, M. S., & Stearns, F. C. (1987). The technical writer's responsibility for safety. *IEEE Transactions on Professional Communication, 30,* 127–132.

Benjamin, A. (1989). *Translation and the nature of philosophy: A new theory of words.* London, UK: Routledge.

Benjamin, W. (2008). The task of the translator. In L. Venuti (Ed.), *The translation studies reader* (pp. 75–84). London, UK: Routledge.

Byrne, J. (2007a). Caveat translator: Understanding the legal consequences of errors in professional translation. *The Journal of Specialised Translation, 7,* 2–24.

Byrne, J. (2007b). The multidimensional challenges of technical translation. *JodyByrne.com.* Retrieved from http://www.jodybyrne.com/144/

Byrne, J. (2009). The coming of age of technical translation. *The Journal of Specialised Translation, 11,* 2–5.

Byrne, J. (2010). *Technical translation: Usability strategies for translating technical documentation.* Dordrecht, The Netherlands: Springer.

Cargile-Cook, K. (2002). Layered literacies: A theoretical frame for technical communication pedagogy. *Technical Communication Quarterly, 11,* 5–29.

Chriss, R. (2006). *Translation as a profession.* Self-published: Lulu.com.

Delisle, J. (1980). *L'Analyse du discours comme methode de traduction.* Ottawa, Canada: University of Ottawa Press.

English, L. K. (1995). Product liability and the plant engineer: An introduction. *IEEE Transactions on Industry Applications, 31,* 1334–1340.

Eubanks, P. (1998). Genre and technical translation: Social, textual, and educational exigence. *Journal of Business and Technical Communication, 12,* 50–70.

FRED, Federal Reserve Economic Data, Federal Reserve Bank of St. Louis: Exports of Goods and Services (EXPGSA); U.S. Department of Commerce: Bureau of Economic Analysis. Retrieved September 21, 2013 from http://research.stlouisfed.org/fred2/series/EXPGSA

Goodwin, J. T. (1996). "Cover your tracks": A case study of genre, rhetoric, and ideology in two psychological reports. *Journal of Business and Technical Communication, 10,* 167–186.

Hackos, J. (1994). *Managing your documentation projects.* New York, NY: John Wiley & Sons.

Hallman, M. (1990). Differentiating technical translation from technical writing. *Technical Communication, 37,* 244–247.

Hammond, M. (1995). A new wind of quality from Europe: Implications of the court case cited by Holz-Mäntäri for the US translation industry. In M. Morris (Ed.), *Translation and the law* (pp. 233–246). Philadelphia, PA: John Benjamins.

Hannah, M. (2011). Legal literacy: Coproducing the law in technical communication. *Technical Communication Quarterly, 20,* 5–24.

Helyar, P. S. (1992). Products liability: Meeting legal standards for adequate instructions. *Journal of Technical Writing and Communication, 22,* 125–147.

Helyar, P. S., & Doudnikoff, G. M. (2003). Walking the labyrinth of multimedia law. *Technical Communication, 50,* 497–504.

Herrington, T. K. (2003). *A legal primer for the digital age.* New York, NY: Pearson.

Hoft, N. L. (1995). *International technical communication: How to export information about high technology.* New York, NY: John Wiley & Sons.

Holz-Mänttäri, J. (1987). Ausgangstext oder Produktspezifikation als beurteilungsgrundlage? Ein gerichtsurteil setzt maßstäbe. *TextconText, 2,* 177–178.

Kastman-Breuch, L. (2002). Thinking critically about technology literacy: Developing a framework to guide computer pedagogy in technical communication. *Technical Communication Quarterly, 11,* 267–288.

Kelly, K., & Jones, J. (2007). Web sites and the law: An avenue for localization. In K. St.Amant (Ed.), *Linguistic and cultural online communication issues in the global age* (pp. 202–213). Hershey, PA: IGI Global.

Lipus, T. (2006). International consumer protection: Writing adequate instructions for global audiences. *Journal of Technical Writing and Communication, 36,* 75–91.

Madden, M. S. (1988). The duty to warn in products liability: Contours and criticism. *Journal of Products Liability, 11,* 103–179.

Markel, M. (2001). *Ethics in technical communication: A critique and synthesis.* Westport, CT : Ablex.

Mayer-Schöneberger, V. (1999). Translation and the law: The view from Europe. *Language International, 11,* 15.

Maylath, B. (2001). Translating user manuals: A surgical equipment company's "quick cut." In D. Boslye (Ed.), *Global contexts: Case studies in international technical communication* (pp. 64–80). Boston, MA: Longman.

McGuire, E. P. (1988). Product-use instructions: How to evaluate them in manufacturer negligence. *Journal of Products Liability, 11,* 203–207.

McKee, H., & Porter, J. (2010). Legal and regulatory issues for technical communicators conducting global Internet research. *Technical Communication, 57,* 282–299.

Morris, M. (Ed.). (1995). *Translation and the law.* Philadelphia, PA: John Benjamins.

Mowat, M. (1998). *Legal liability for information provision.* London, UK: ASLIB

Muegge, U. (2005). *Translation contract. A standards-based model solution.* Bloomington, IN: AuthorHouse.

Nagelhout, E. (1999). Pre-professional practices in the technical writing classroom: Promoting multiple practices through research. *Technical Communication Quarterly, 8,* 285–300.

Pander Maat, H., & Lentz, L. (2011). Using sorting data to evaluate text structure: An evidence-based proposal for restructuring patient information leaflets. *Technical Communication, 58*(3), 218–237.

Pommer, S. (2008). No creativity in legal translation? *Babel, 54*(4), 355–368.

Révy von Belvárd, K. (1997). The legal importance of technical translations. *Technology, Law and Insurance, 2,* 191–194.

Ross, K. (1981). Legal and practical considerations for the creation of warning labels and instruction books. *Journal of Products Liability, 4,* 29–45.

Sanderlin, S. (1988). Preparing instruction manuals for non-English readers. *Technical Communication, 35,* 96–100.

Schoff, G. H., & Robinson, P. A. (1984). *Writing and designing operator manuals.* Belmont, CA: Lifetime Learning.

Selber, S. (2004). *Multiliteracies for a digital age.* Carbondale, IL: Southern Illinois University Press.

Slattery, S. (2007). Undistributing work through writing: How technical writers manage texts in complex information environments. *Technical Communication Quarterly, 16,* 311–326.

Smith, H. (1990). Technical communications and the law: Product liability and safety labels. *Journal of Technical Writing and Communication, 20,* 307–319.

Smith, H. T., & Shirk, H. N. (1996). The perils of defective documentation: Preparing business and technical communicators to avoid products liability. *Journal of Business and Technical Communication, 10,* 187–202.

Spilka, R. (Ed.). (2010). *Digital literacies for technical communication: 21st century theory and practice.* New York, NY: Routledge.

St.Amant, K., & Rife, M. C. (2010). Legal issues in global contexts: Reconsidering content in an age of globalization. *Technical Communication, 57,* 249–250.

Stejskal, J. (2006). Quality assessment in translation. *ATA Chron*icle, 35, 12–16.

St. Germaine-McDaniel, N. (2010). Technical communication in the health fields: Executive Order 13166 and its impact on translation and localization. *Technical Communication,* 57, 251–265.

St. Louis Federal Reserve Data Service (FRED). (n.d.). Retrieved from http://research.stlouisfed.org/fred2/graph/?id=EXPGSA

Strate, L., & Swerdlow, S. (1987). The maze of the law: How technical communicators can research and understand legal matters. *IEEE Transactions on Professional Communication, 30,* 136–148.

Taylor, T. (1996). Legal literacy and the undergraduate curriculum. *Journal of Business and Technical Communication, 10,* 239–250.

Translation errors cause lost revenue in 80 percent of global firms. (2007). *BusinessWire.* Retrieved from http://www.businesswire.com/news/home/20070330005539/en/Translation-Errors-Lost-Revenue-80-Percent-Global

Ulijn, J. (1996). Translating the culture of technical documents: Some experimental evidence. In D. Andrews (Ed.), *International dimensions of technical communication* (pp. 69–87). Arlington, VA: Society for Technical Communication.

Velotta, C. (1987). Safety labels: What to put in them, how to write them, and where to place them. *IEEE Transactions on Professional Communication, 30,* 121–126.

Venuti, L. (1995). *The translator's invisibility: A history of translation.* London, UK/New York, NY: Routledge.

Venuti, L. (1998). *The scandals of translation: Towards an ethics of difference.* London, UK/New York, NY: Routledge.

Vignali, R. M. (1995). Foreign language warnings and the duty to warn. *Risk Management, 42,* 83–91.

Weiss, T. (1995). Translation in a borderless world. *Technical Communication Quarterly, 4,* 407–423.

Weiss, T. (1997). Reading culture: Professional communication as translation. *Journal of Business and Technical Communication, 11,* 321–338.

http://dx.doi.org/10.2190/LIIC6

CHAPTER 6

Adapting to the Changing Legal Context: How Executive Order 13166 is Shaping Translation and Localization in Health Care

Nicole St. Germaine

Each day, hospitals and clinics around the United States encounter limited English proficient (LEP) patients: patients whom the United States Census Bureau defines as "speak[ing] English less than 'very well'" (American Medical Association [AMA], 2006, p. 1). And how many prospective LEP patients are there in the United States? According to the Census Bureau's 2009 American Community Survey, 19.6% of Americans over the age of 5 speak a language other than English at home. Of that population of approximately 55 million individuals, 43.8% (i.e., 24 million people) speak English "less than very well" (Languages spoken at home, 2009). The breakdown of those who speak English "less than very well" is as follows:

- 46.9% of Spanish speakers,
- 48.5% of Asian and Pacific Islanders,
- 32.8% of Indo-European language speakers, speak English "not well" or "not at all." (Languages spoken at home, 2009)

These numbers imply that 10% of Americans over the age of 5 do not speak English "very well," and that is a startling number. These populations, moreover, are distributed unevenly across the United States, with most of the individuals who speak less than "very well" concentrated in the western states, the Southwest, Florida, and the Northeast. Generally speaking, this number means that 1 in 10 patients in U.S. hospitals need language assistance of some sort in order to obtain appropriate health care.

Additionally, a 2006 survey conducted by the Health Research and Educational Trust found that 80% of hospitals encounter LEP patients "frequently," which was defined as at least monthly (Hasnain-Wynia, Yonek, Pierce, Kang, & Hedges Greising, 2006, p. 2). This number represents a very large population of LEP patients who need translation (or written language services) and interpretation (or oral language services) in a healthcare setting (p. 29).

The topic of medical translation and interpretation is an issue not only in the United States; worldwide, there are an estimated 191 million immigrants, and the immigration rates have more than doubled in the past 50 years around the globe (Shah, 2008). In fact, the United States accounts for only 20% of worldwide immigrants, while Europe absorbs 33% of new immigrants globally. Even if only a fraction of this number requires language-related help, the resulting population in need of such services is very large. This situation, in turn, represents what many consider to be a global health information crisis (Shah, 2008). Within this context, the European Union is facing its own unique challenges, particularly in the regulation of documentation for medical devices. The European Union makes up 30% of the world's market for medical devices, and it continually encounters problems balancing its need to meet translation-related regulations against the cost and time required to do so (Regulatory language requirements, 2005).

Research indicates that many organizations are attempting to accommodate these populations by offering a greater number of translated materials for their clients; many organizations in the United States, such as the American College of Obstetricians and Gynecologists (ACOG) and the Centers for Disease Control (CDC), have been translating and localizing health materials in greater numbers after the passage of Executive Order 13166 in 1990 (St. Germaine, 2009; St. Germaine-Madison, 2007). This global need for more translation and localization in medical contexts has led to an increased involvement of technical communicators as writers and as localization experts in the health-related fields in every developed country. Within this context, understanding the linguistic needs of the populations being served is central to successful communication and to successful legal compliance with various national and international laws.

In the United States, understanding the needs of LEP populations in a healthcare setting is still incomplete, for until now, the focus within the healthcare sector has largely been industry-oriented. Executive Order 13166, however, has served as an impetus for technical communicators in the United States to study and to serve LEP populations in a variety of medical contexts, as well as to examine localization needs within these settings.

Previous works on Executive Order 13166 and its impact on translation and localization have sought to understand how technical communicators working in the healthcare industry can further the effort to make hospitals and clinics more compliant with EO 13166. The need to bring these healthcare settings into compliance with EO 13166 remains a very pressing, yet daunting, task. As of yet, few hospitals or clinics can claim that all of their essential medical information is available through translation or interpretation all of the time. This situation was noted by the U.S. Attorney General Eric H. Holder, Jr. in a memo sent to the heads of all Governmental Components on June 28, 2010. Entitled "Language Access Obligations under Executive Order 13166," the memo reminded these department heads of their obligations to and the tenets of EO 13166. As Holder (2010) explains, "A decade later, we have much work ahead of us to transform policy into practice and make the Executive Order's mandate a reality." This news comes as no surprise to either technical communicators working in the healthcare fields or to healthcare professionals who work with LEP patients each day.

In this chapter, I examine prospective initial questions, issues, and problems that seem to affect the conversation technical communicators might encounter about what the day-to-day realities of healthcare providers are when working with LEP patients within the context of EO 13166. Specifically, I present the results of a small-scale pilot study done to uncover prospective areas for future research on this topic area. This pilot involved in-depth interviews with four health professionals who come into regular contact with LEP patients. The purpose of these interviews was to gain initial insights into how translated materials and interpretation work (or fail to work) in the context of trying to communicate with patients.

In conducting this pilot study, I spoke with the following individuals:

- An RN working in an ICU unit in an urban hospital in Detroit
- An oxygen technician working in Boston
- An RN working in a labor and delivery unit in a smaller urban setting in West Michigan
- A receptionist working in a busy clinic in Albuquerque

I chose these areas not only for convenience, but also because they represent some of the most LEP-populated areas in the country as well as individuals working in the healthcare settings who generally have the highest chances for encounters with LEP populations.

OVERVIEW OF EXECUTIVE ORDER 13166

In August of 2000, President Bill Clinton issued *Executive Order 13166: Improving Access to Services for Persons with Limited English Proficiency* (http://www.usdoj.gov/crt/cor/Pubs/eolep.htm). The intent of this order was twofold. First, the order was designed to better enforce Title VI of the Civil Rights Act of 1965. Title VI "prohibits recipients of federal financial assistance from discriminating based on national origin by, among other things, failing to provide meaningful access to individuals who are limited English proficient" (LEP,

2011). This order, in effect, works to "ensure persons with limited English proficiency get meaningful access to federally funded programs" (Wong, 2004, p. 2). And why is this an issue? The answer is that many agencies, public or private, receive some sort of federal assistance.

In the case of health care, if a clinic, practice, or hospital accepts Medicaid or Medicare, this constitutes federal assistance. This situation is even true if only one area of the hospital or clinic accepts federal funding (Schroeder, 2002, p. 2). As a result, entities now affected by EO 13166 include most "hospitals, doctor's offices, nursing homes, managed care organizations, state Medicaid agencies, home health agencies, health service providers, and social service organizations" (Sampson, 2006, p. 134).

Because of this mandate, hospitals, clinics, and other healthcare providers across the United States have been scrambling to provide patients who are nonnative English speakers with translated materials and interpreter assistance. Even a decade after the order was issued, many of these institutions are still working to increase their compliance with it (Barrett, Dyer, & Westpheling, 2008). This situation means technical writers and localization experts in the healthcare fields will have work for years to come to even get close to making current programs, clinics, and hospitals compliant with EO 13166.

In terms of the translations, EO 13166 works to remedy the problems caused by language barriers by requiring that all medical documentation be translated as part of a "vital" service (LEP, 2011). What is considered to be vital can vary, and which documents are defined as vital falls, in part, under the jurisdiction of each agency. However, the FAQ page for the order (see http://www.lep.gov/faqs/faqs.html) labels "vital documents" as those that contain

> [I]nformation that is critical for obtaining federal services and/or benefits, or is required by law. Vital documents include, for example: applications; consent and complaint forms; notices of rights and disciplinary action; notices advising LEP persons of the availability of free language assistance; prison rule books; and written tests that do not assess English language competency, but rather competency for a particular license, job, or skill for which English competency is not required; and letters or notices that require a response from a beneficiary or client. (FAQ 9)

Interestingly, the FAQ page itself acknowledges "It may sometimes be difficult to draw a distinction between vital and non-vital documents, particularly when considering outreach or other documents designed to raise awareness of rights or services" (FAQ 9).

Further, many agencies and hospitals wishing to be in compliance with the executive order have noted problems with informed consent. The question, in essence, is "How can an agency or healthcare setting be sure that a patient truly understands the procedure he or she is about to undergo?" Many, such as the

National Health Law Program, recommend the use of printed patient materials to elaborate on an interpreter's explanation or to substitute for an interpreter if it is necessary (Sampson, 2006). Therefore, materials such as the patient information pamphlets written by medical writers and technical communicators and issued by ACOG, which are handed out at OB/GYN offices, can become vital tools for ensuring compliance with the order. As a result, while medical interpretation might be the preferred format, the translated pamphlets are necessary to fulfill legal requirements or to serve in circumstances in which appropriate medical interpretation is not available.

Failing to meet these interpretation and nondiscrimination standards may have potentially severe consequences for clinics, hospitals, and programs judged not in compliance. In her article "Meeting the Needs of Patients with Limited English Proficiency" (2001), Colleen M. Roberts explains that the government can terminate medical contracts with any physicians or agencies that fail to provide some sort of language assistance, and thus do not comply with EO 13166. Further, Roberts points out there is a legal precedent set by the courts:

> Health care providers not providing language assistance are negligent in facilitating clear communication and are ultimately obstructing informed consent, thus putting themselves at fault for malpractice suits where language services would have been appropriate. (p. 71)

Accordingly, if an individual or organization issues a translation-related complaint to the Department of Health and Human Services Office for Civil Rights, the complaint will be investigated. If the healthcare organization or office is found guilty, it might lose its right to receive federal funds such as Medicare, Medicaid, or the State Children's Health Insurance Program (SCHIP) (Sampson, 2006, p. 134).

A survey of the literature reveals that many organizations have turned to technical communicators, including experts in localization and writing for translation, in order to be compliant with EO 13166 (see Perkins, 2003; Sampson, 2006). The overarching idea is that everything from the standards that the organization develops to meet EO 13166 to the actual documentation in question are developed in conjunction with localization experts. Technical communicators, in turn, work with localizers who help ensure that the documents being produced meet the cultural preferences and the linguistic needs (in terms of style and content) for the target audience. As a result, technical writers and localization experts have become the lynchpin on which the success of these document revisions turn.

In order to effectively assist healthcare providers with their information needs, it is important to examine the realities and limitations with which these professionals work. EO 13166 is, admittedly, a best-case scenario in which each and every patient can walk into any healthcare setting nationwide and get the language assistance he or she needs. Even 10 years later, however, we are far from reaching that ideal for a variety of reasons. These limitations often have to do with the difficulty of setting up a scenario favorable to universal language assistance in busy healthcare settings. Such environments would include an emergency room or even mobile healthcare settings, like a technician's visit done to set up durable medical equipment in a patient's home.

METHOD

In order to investigate the real-life challenges healthcare practitioners face with working with LEP populations, I conducted a pilot study composed of 16 interviews with healthcare workers across the country. In a number of cases, however, the practitioners wanted to stop the interview in midprocess or later denied me permission to use their interview because they realized that their healthcare setting was not compliant with EO 13166. For this reason, these interview subjects feared being fired or placing their workplace in danger of a lawsuit. As a result, I was able to use only 4 of the original 16 interviews. And for these four cases, the interviewees asked that I name only their industry and city or region rather than name their employers specifically in order to avoid reprisals or negative consequences for their employers, and I, in turn, used pseudonyms when conveying the results of these interviews. While the resulting information was greatly limited, it did provide interesting initial insights into this topic, and these insights can, I believe, serve as a foundation others can use to further investigate key questions, concepts, and trends that emerged.

Interview Participants

Four individuals agreed to participate in the interview and have their information used for the chapter. Because of the small sample size, the results of this sample cannot be generalized to a larger population of healthcare professionals. However, this small-scale ethnography serves as an entry point, or a pilot examination, that can be used to play/develop future research on this topic.

Each participant works in a different area of the country and has a different job description (see Table 1).

Questionnaire

Once the interviewees agreed to participate in the study, they were emailed a short survey to complete. The survey explained the intent of this project, as well as offered a short definition of Executive Order 13166. The survey questions were the following:

Table 1. Interview Participants and Related Backgrounds

Interviewee	Background (workplace and region)
Gregory	An RN in an ICU Unit in an inner city hospital in Detroit, Michigan. Gregory works in direct patient care in a critical care setting. Detroit is similar to many of the large northern industrial cities in that it has a large population of residents who were born in other nations. Detroit, however, is unique in that over 50% of the region's immigrants hail from the Middle East or Asia. Nationwide, over 67% of the immigrants living in the United States originate from American countries. By contrast, only 21% of the immigrants living in the Detroit region come from the Americas (Armstrong, 2009). Detroit is therefore an interesting area to examine in terms of *EO 13166* compliance, for most health translation efforts nationally are focused on serving the Spanish-speaking populations.
Melinda	A reception supervisor in a busy clinic in Albuquerque, New Mexico. Melinda's main duties include answering telephones and scheduling appointments for a multi-specialty medical group. Melinda's workplace setting in Albuquerque is similar to much of the American Southwest in that it has a large Spanish-speaking population. What makes Melinda's case particularly interesting is that, in 2000 (the most recent year for which specific language data is available for this region), 36.7% of New Mexico's population spoke a language other than English, with the majority of the LEP population speaking Spanish (New Mexico Languages, 2010). Therefore, New Mexico is an interesting area to focus on for *EO 13166* compliance due to the high need of language assistance in the nation's most widely spoken non-English language.
Julia	An RN in a labor and delivery unit in a hospital in West Michigan. Julia's main duties involve assisting with labor and delivery as well as direct patient care after delivery. Often, rural areas such as West Michigan get ignored when language assistance data is collected because it is thought not as much need for language assistance exists in these regions. However, many agricultural areas, like West Michigan, host a large number of migrant farm workers (most of which come from the Americas) each summer (Miklavcic, 2004).
John	An oxygen technician working for a national durable medical equipment company in Boston, MA. John's main job duties include setting up oxygen equipment and other durable medical equipment in homes and hospitals, repairing and troubleshooting the equipment, and educating the patient and the patient's family about how to use and maintain the equipment. This work places him in an interesting context. The Massachusetts immigrant population is at a 50-year high, representing 14% of the labor force, and the Boston area is home to immigrants from areas as diverse as Haiti, Vietnam, the Azores, Brazil, The Dominican Republic, and China (Abraham, 2006). Similar to Detroit, the diverse populations located in Boston (as well as many other large cities in the Northeast) represent a significant challenge for *EO 13166* compliance.

Demographics

1. Name (as you would like it to appear in the chapter):
2. Occupation:
3. Major job duties:
4. Employer (you can omit if you prefer or request that I don't use the name):
5. City and state (or region, if you prefer to remain more anonymous):

Interview Questions (please be as detailed as possible)

1. Before I contacted you, had you heard of Executive Order 13166 or the fact that essential health care information must be translated or interpreted for your patients?
2. How often, would you estimate, that you come into contact with non-English speaking or limited English-proficient (LEP) patients in your occupation?
3. What languages do most of the people you encounter speak (other than English)?
4. How does your workplace accommodate patients who are non- or limited English speaking?
5. If translations of materials are available, what forms, documents, etc. are translated?
6. Are there forms or other documents that do not have a translation available that (in your opinion) should have a translation?
7. Are interpreters available for non-English speaking or LEP patients in your workplace? (in person or telephone interpreters?) If so, how well does using an interpreter work for these patients?
8. In your experience, what are the greatest difficulties in your position with working with non-English speaking or LEP populations? Can you give examples?
9. Any other comments?

While the results of this small-scale pilot research cannot be generalized, they can help identify certain issues that merit further investigation in relation to the implementation of EO 13166.

INTERVIEW RESULTS

What follows is a summary of the initial pilot data I was able to collect through this small-scale, exploratory research:

Responses to Question 1: Before I contacted you, had you heard of Executive Order 13166 or the fact that essential health care information must be translated or interpreted for your patients?

Interestingly, of the 16 interview participants, including the 12 who later decided not to have their detailed responses become part of the chapter, only Julia, the labor and delivery nurse, and John, the oxygen technician, knew of EO 13166 specifically. Both stated that they had undergone training from their employers about how to help LEP patients and what the law entailed. Gregory, the ICU RN, stated that he didn't know the name of the law or what the contents of the order were. He did, however, note that offering translations and interpreters for LEP patients was the "standard of care" that was expected in his hospital. As a supervisor of reception at an urban clinic, Melinda's experiences were different, for she does not often come into contact with "vital" medical documents such as HIPPA forms or treatment consent forms, which might explain why she had not received training on the topic.

The general lack of awareness about this specific order illustrates the struggle many healthcare settings have with compliance with the Order. Not only must an interpreter and specific translations be made available, but the staff must also be trained to use these materials. As of this writing, no state offers standard training for working with LEP patients; rather, it has been left up to individual clinics and healthcare settings to address such factors (Barnett et al., 2008). This factor might explain the disparities in the amount of information that the healthcare professionals had.

Responses to Question 2: How often, would you estimate, that you come into contact with non-English speaking or limited English-proficient (LEP) patients in your occupation?

Bill, Melinda, and Gregory all mentioned that they came into contact with LEP patients daily, if not multiple times each day. Citing the ethnic enclaves that are characteristic to some areas in Boston, John stated that, in some cases, entire clinics would be largely LEP: "There was a time we received a contract from a clinic that was 90% non-English speaking. The other 10% spoke very little English. They were all receiving nebulizers for COPD treatment."

As an RN in a regional hospital in West Michigan, Julia's situation was more unique in that her contact with LEP patients varied according to the seasonal fluctuation of migrant farmworker populations in the area. Julia stated she anticipated an increase in the frequency of LEP patients she sees because the small community hospital that mainly services those areas would no longer offer routine OB care. In her 9 years in working in OB care, she had seen an average of 1–2 non-English-speaking patients per week. West Michigan is similar to many agricultural regions in the northern states in that it has few LEP residents in the winter, but many more in the summer because of the migrant farmworker population (Miklavcic, 2004).

The fact that healthcare professionals working in diverse settings across the country all regularly come into contact with LEP populations is not surprising, given the census data that indicates that 1 in 10 Americans does not speak English well or at all. Although the LEP population tends to be concentrated in the South and Southwest and in urban centers, they are also frequently found in agricultural settings as well as smaller towns across the country. (Such distribution is often the result of a range of economic and social factors that push individuals north and east in search of jobs; Camarota & Jensenius, 2009.) These factors mean that healthcare workers and technical writers working in the healthcare fields across the country will increasingly come into contact with LEP populations. As a result, even healthcare settings that have largely ignored EO 13166 because they have not come into regular contact with LEP patients will likely have to increase their efforts to meet the needs of these patients.

Responses to Question 3: What languages do most of the people you encounter speak (other than English)?

All interview participants named Spanish as the primary language (other than English) they encounter, and the rest of the results varied, with Russian dialects, Vietnamese, Chinese, Urdu, Hindi, Arabic, Haitian Creole, Portuguese (Brazilian dialect), and Québécois French, and American Sign Language also mentioned.

Responses to Question 4: How does your workplace accommodate patients who are non- or limited-English speaking?

Even though every interview participant noted Spanish as the dominant second language they encountered (after English), interestingly, none of the participants mentioned Spanish as problematic, for systems tended to be in place for handling Spanish-speaking patients. For instance, John relied on his own limited knowledge of medical Spanish as well as Spanish-language brochures and forms that served to explain the operation of the medical equipment to Spanish-speaking patients. Melinda, Gregory, and Julia relied on colleagues who spoke Spanish to handle Spanish-speaking patients. This is not to say that Spanish-language assistance is always readily available, particularly in areas without a high Spanish-speaking population. However, it does indicate that many hospitals and clinics are becoming increasingly compliant with EO 13166 in terms of providing language assistance to Spanish-speakers.

In these cases, the problem seemed to lie with the other, non-Spanish and non-English languages. Here, none of the interview participants cited any knowledge and limited or nonexistent ability for co-workers to help such patients. John in particular ran into trouble within the context of his work as an oxygen technician (a job that often required him to meet with patients in their homes). In many cases, if the patient or the patient's family couldn't speak English or Spanish, John was required to use his cell phone to call a co-worker who could speak that other language. This approach served as a cost-saving measure in that the company could rely on its employees to help one another rather than pay for an interpretation service. However, this reliance on employees for translation could also be problematic because the employee who spoke the patient's language was often not available. In instances in which no one was available to interpret for the patient, John was forced to set up the equipment and leave without being able to tell the patient where he was going or when the patient could expect further instruction about the use of the medical equipment he had delivered. In such situations, John would return later when an interpreter was available to help explain to the patients how to use the equipment and how to fill out the related insurance paperwork. When John did return (up to 5 hours later), he would often find a very upset patient and family who had imagined that they had been left to figure out how to use the equipment on their own.

The situation that John describes is technically compliant with EO 13166, for the Order only mentions that interpretation or translation service must be available. It does not comment on the timeliness of its delivery in a noncritical situation, such as the assembly of a nebulizer for a patient with chronic obstructive pulmonary disease (COPD).

Further, the use of an interpreter often becomes impractical for healthcare workers in emergency settings. Gregory, the RN working in an ICU unit, stated that in the ICU, many patients are unresponsive or have a ventilator with an oral endotracheal tube (OET or breathing tube) in their mouth and are therefore unable to speak with an interpreter. Moreover, even if the patient were able to speak, there are no telephones at the bedsides in ICU rooms. As a result, nurses in Gregory's ICU unit had to rely on staff or on a family member to interpret for the patient. In such scenarios in which a patient is unconscious or is unable to respond, or if time is of the essence, healthcare providers often are forced to forgo the interpretation service. They must instead act quickly for the benefit of the patient. These scenarios create a sticking point for compliance with EO 13166, which makes no allowance for emergency settings.

Other times, an interpreter might not be deemed necessary. In the case of giving patients medicine, fact sheets in a variety of languages might be available. Gregory, for example, encountered a situation in which the patient, a Spanish-speaker, had an additional question about the purpose of a particular medication. In cases such as these, healthcare providers in Gregory's hospital rely on basic machine translation to yield a simple response. As Gregory explained, "If we know the patient's language, a quick Google search will yield translations of some basic medical statements."

Machine translation, however, is notoriously unreliable. Moreover, it has been cited (along with faulty human translation) with causing inappropriate dosing in medicine. The problem is that medical translations necessitate an understanding of the associated field in which translation takes place. For example, to accurately translate technical terms, concepts, or processes in cardiology, toxicology, or pharmacology, one needs to know and to understand such concepts in order to correctly convey information in another language (Medical Translation, n.d.). Often the translator needs to understand the culture for which he or she is translating in order to convey information in appropriate and effective ways. For instance, the South Cove Community Health Center in Boston found that Chinese-language instructions for Chinese patients had to be localized because the volume of fluid Chinese patients associated with "teaspoons" was much larger than thc volumc of fluid most Anglo Americans associate with teaspoon (U.S. Department of Health and Human Services, 2001, p. 18). As machine translation does not account for such cultural or contextual information, it thus often yields inaccuracies that might prove deadly for patients.

For simple medical statements, machine translation might prove to be a cost-saver, but it must be used with extreme caution. Even statements that seem straightforward can be problematic when the cultural context is considered. For instance, the statement "You have tested positive for HIV," when translated into Spanish yields "Usted ha probado positiva para el VIH" with Yahoo's Babel Fish machine translator. The problem here is that in Spanish, "positive" indicates a good result rather than indicating that one has HIV.

Responses to Question 5: If translations of materials are available, what forms, documents, etc. are translated?

John, Gregory, and Julia mentioned that all of their consent forms and other "vital" forms are available in Spanish. Gregory also cited the availability of materials translated into French and Arabic, as one might expect in Detroit with its large Arabic-speaking population and its contact with French-speaking Canada. Julia's response was the most comprehensive and covered the largest range of available documents:

> In hard copy on our unit, we have necessary forms translated into Spanish. These include consents for treatment, consents for surgery and anesthesia, consents for the receipt of blood products, and consents for vaccinations. We also have educational materials in Spanish, including breastfeeding and bottle feeding instructions, information on newborn screening and hearing screening, hepatitis B vaccine information, and basic infant care instructions. The computer allows us to handle printing medication and discharge instructions, as well as information on health conditions.

Most notably, materials in languages other than Spanish were available in Julia's hospital: "Through our computer system, we have the ability to print discharge instructions, medication instructions, and other educational materials in a large number of languages as well." Admirable as such a system is, it is perhaps easier to provide this breadth of languages because of the more limited range of conditions that Julia's unit deals with in the context of labor and delivery. For Gregory's situation in the ICU, where a virtually limitless number of conditions are seen, it would be far more difficult (and perhaps impractical) to provide patient information in many languages instantly. Similarly, although John mostly works with lung conditions, it would be difficult for him to keep all of the possible materials on hand as he travels from home to home or to the hospital.

Oddly, Melinda's office in Albuquerque had forms available only in English. Perhaps this is because most of the forms they have, such as insurance forms, are not necessarily considered "vital" documents under the definition set forth by EO 13166. Therefore, such materials might not be translated to guarantee compliance with the Order.

Responses to Question 6: Are there forms or other documents that do not have a translation available that (in your opinion) should have a translation?

Of the interviewees, Melinda was satisfied with the language support in the office and felt that the assistance of bilingual office staff eliminated the need for forms in other languages. Gregory, in turn, felt that by and large written forms were impractical in an ICU setting (favoring the assistance of interpreters instead). The other interviewees, however, expressed concerns with the availability of forms. John in particular was dissatisfied with the lack of availability of HIPPA forms in languages other than Spanish. His concern was that patients who spoke other languages (a very common scenario in Boston) were not being properly notified of their rights:

> This is almost certainly violating the law, and I think if we could just keep a few of these consent-to-treat forms on hand, in say, Haitian Creole, Portuguese, and Vietnamese, it would cover the majority of the LEP patients that I assist.

Julia was satisfied with her ability to obtain consent from patients, but not with the availability of specific consent forms. She was particularly concerned because, when the interpreter assists an LEP patient with a consent form, she must go over the form word for word with the patient while the attending RN charts the entire process to ensure that the patient has been read her rights adequately. Julia stated that this process could be greatly simplified if she could just hand the patient a form in the correct language.

The ready availability of educational materials, but the related lack of translated consent forms, makes one wonder why these documents were excluded from the system Julia mentioned earlier. This situation, coupled with John's lack of HIPPA forms in languages other than English or Spanish, is a cause for concern. (These forms are standardized, and it would seem to be relatively simple to obtain translations of these forms in a variety of languages.) Of course, the responses of two individuals are not enough to generalize about the availability of such forms in translation nationwide. They do, however, indicate an area wherein technical communicators might wish to do more research to determine which kinds of health-related materials are being translated, where, and why.

Responses to Question 7: Are interpreters available for non-English speaking or LEP patients in your workplace? (In-person or telephone interpreters?) If so, how well does using an interpreter work for these patients?

All respondents mentioned the use of telephone interpreters to address language issues, Few, however, were completely satisfied with the process. Julia was mostly pleased with the efficiency of the telephone interpreter process, except in emergency situations. As she explained, "In emergent situations such as a precipitous delivery or emergency cesarean section, the phone interpreter system is not always a viable option for a patient in extreme pain or a patient who is pushing" (a sentiment echoed by Gregory in noting similar difficulties with the ICU).

Again, while the sample size of this pilot study was small, it does help identify certain areas that might benefit from future research. These include

- How to address issues of translation in emergency situations when aspects of time and speed make the use of phone interpretation highly impractical.
- How text-only translations pose problems for patients who are in extreme pain or who are not conscious or able to understand what is being said because of their condition.

In such cases, the seemingly practical solution would be to use an on-site interpreter, yet doing so poses problems in terms of cost and availability. Even when on-site interpreters are available, they often get in the way in emergency situations, as Gregory explained, "The only time I've had an interpreter at the bedside was for a deaf patient who had someone signing to him. It was fine, but the interpreter took up a lot of space in a tiny (about 10' × 12') room." Nonetheless, such situations will likely become more common over time. Thus, more research is needed on how to address such situations effectively.

Responses to Research Question 8: In your experience, what are the greatest difficulties in your position with working with non-English speaking or LEP populations? Can you give examples?

The problematic issue most frequently mentioned by interview participants was the frustration with not being able to speak Spanish. Julia in particular appreciated the phone interpreter. At the same time, she felt that using one interfered with her standard of care: "Connecting with patients via a caring attitude and body language is universal, but using a phone to communicate inserts a barrier not present with a common language."

Similarly, John mentioned the importance of learning basic medical Spanish. He stated that he spoke "a small amount of Spanish, which I learned in high school, as well as some medical terms that one of co-workers (from the Dominican Republic), gave us a quick 'workshop on pronunciation.' Every little bit helped." He added that he would encourage all healthcare practitioners to learn some basic medical Spanish. As he noted, even though it won't help with every patient, he found that in speaking Spanish and English, he could communicate with the patient at least 90% of the time.

Another basic concern was the literacy level of the patient. Gregory in particular had concerns with both the textual translations and the interpreter because of the frequency with which he encountered patients who have either limited reading ability or limited experience with medical terminology. Gregory's anxiety centered on the fact that limited literacy levels in the Detroit area confound the situation. However, he argued that oversimplifying medical documents might backfire, for such materials might not fully convey the intended or the desired message. As Gregory explained, "For patients who are from countries where literacy is low, printed materials may be of little to no value."

Another problem that healthcare professionals encounter, particularly in the southwestern states such as Texas and New Mexico, is the use of local dialects of Spanish. These dialects often make the use of "neutral" or "universal" Spanish inappropriate, for many Spanish-speaking patients might not be literate in this standard Spanish dialect. In my own research, I have found this to be true of Spanish-speaking populations native to Texas (or *Tejanos*). In a number of cases, the "standard" Spanish dialect being used was not as widely understood by a given target population because of differences in vocabulary and syntax between the *Tejano* dialect and the standard Spanish (St. Germaine-Madison, 2007).

Melinda encountered similar problems in New Mexico:

> The "Spanglish" spoken by the vast majority of New Mexicans is not the same as the Mexican language. That is the main reason why we hire a translator from an outside firm or have the patient's insurance company send someone over to translate.

As Melinda points out, the use of a dialect can complicate matters and raise fears of liability if the patient doesn't understand the interpreter.

RECOMMENDATIONS FOR TECHNICAL AND HEALTH COMMUNICATORS

The legal, cultural, and linguistic complexity of the situation and the wide variety of healthcare settings that fall under the jurisdiction of Executive Order 13166 make it difficult for technical and health communicators to know where to start. Additionally, technical communicators developing medical materials often have little or no say over what gets localized or translated and into which languages. Similarly, they generally have little say in how translations will be distributed. So what can one do to address the current U.S. legal situation created by EO 13166?

Unfortunately, the small size of the pilot study reported here prevents any generalizations on this topic. The study does, however, raise the possibility that healthcare professionals might not have a full understanding of the range of cultural and linguistic factors that need to be considered and addressed within the context of current legal frameworks. Thus, as noted previously, more research in this area is needed—and needed soon—so individuals working in health-related contexts can better address legal issues affecting medical communication in the United States. In the interim, the following general recommendations might help to address this legal context until more research on this topic can be conducted.

Recommendation 1: Consider the Level of Literacy and the Textual Style of the Local Population

When preparing the document for translation, try to match the literacy level of the translated document to the general reading ability of the related target population. As literacy rates among LEP patients can fluctuate widely, texts written at the "standard 4th–6th grade reading level" might be problematic for some patients, even if the information is translated and localized perfectly otherwise.

Similarly, make the content of the information match that of the culture's expectations. One strategy for making the content culturally appropriate (i.e., understandable and socially acceptable) is to conduct research on the communication expectations of different cultural groups (see, for example, Hall, 1976; Hofstede, 2001). Information on the ways in which cultures differ is available, and as the work of Tebeaux (1999) and others suggest, highly applicable to practical writing strategies for cross-cultural contexts.

Recommendation 2: Research the Cultural Preferences for Uses of Graphics and Colors in Medical Documents and Consider These Factors When Creating Materials for Patients From Other Cultures

In some cases, such as China, the United States, and Mexico, there is specific research available on the preferred format of graphics in various contexts.

However, for other cultures (e.g., the Haitian population John regularly encounters in Boston), more research is needed. Technical communicators could try to address this situation through using focus groups, interviews, and other interactions with members of these populations.

Recommendation 3: Use Focus Groups to Test the Documentation

Many organizations agree that focus groups are an effective and a cost-efficient way to localize healthcare information for a particular patient population. The nature and scope of the focus group will vary depending on how widely distributed the information will be and how many target cultures one must address. Many organizations, such as the Community Health Education Center of the Massachusetts Department of Public Health, conduct smaller, locally based focus groups to obtain their information coming from the hospital or clinic. Other, larger organizations, such as the CDC, conduct several focus groups across the country to ensure that they are capturing a wider array of opinions for information that will be published nationally (Ogilvy, 2005; U.S. Department of Health and Human Services, 2001, p. 14). The use of such focus groups can also provide valuable information about a particular culture's preferences for text and graphics when other research is not available through research.

Recommendation 4: Consider the Content, Not Just the Style and Format

The localization of textual healthcare information is a sensitive topic with much at stake for both the patient and for the hospitals and clinics involved. Localizing this information goes beyond a simple consideration of language, style, and graphics. Rather, the information also must match the expectations members of a culture have for the information being presented. This includes the topics that can be covered and the culture's view of health and health care in general (Callister & Birkhead, 2002; Forslund, 1996; Kreps & Kunimoto, 1994). Again, if this information is not known or is not clearly understood, conducting focus groups consisting of members of the target population can be an inexpensive and highly effective way to determine how best to approach a health topic with a particular population.

CONCLUSION

Many, if not most, U.S. healthcare settings are still struggling to become compliant with the tenets of EO 13166. The variety of settings, the condition of the patient, and the availability of funding for translations can thus cause problems with ensuring that each patient receives all of the information that he or she needs. Further complicating these matters is that hospital or organizational administrative boards often make the decision about what gets translated or what merits the use of an interpreter—and do so without input from technical or health communicators involved in such processes. Technical communicators can, however, influence the quality of the text that goes to translation. Even simple measures such as testing documents with focus groups can help effect differences in a patient's understanding of their rights or their condition and can help a hospital or a clinic become compliant with EO 13166.

REFERENCES

Abraham, Y. (2006). Immigrant labor force booming. *The Boston Globe Online.* Retrieved from http://www.boston.com/news/local/massachusetts/articles/2005/06/19/immigrant_labor_force_booming/

American Medical Association (AMA). (2006). *Office guide to communicating with limited English proficient patients.* Chicago IL: AMA.

Armstrong, M. (2009). By the numbers: The Detroit region is one of the most diverse in the nation. *Detroiter Online.* Retrieved from http://findarticles.com/p/articles/mi_hb4831/is_2_99/ai_n29457178/

Barrett, S., Dyer, C., & Westpheling, K. (2008). *Language access: Understanding the barriers and challenges in primary care settings.* McLean, VA: Association of Clinicians for the Underserved.

Callister, L. C., & Birkhead, A. (2002). Acculturation and perinatal outcomes in Mexican immigrant women: An integrative review. *Journal of Perinatal and Neonatal Nursing, 16*(3), 22–38.

Carmarota, S., & Jensenius, K. (2009). A shifting tide: Recent trends in the illegal immigrant population. *Center for Immigration Studies.* Retrieved from http://www.cis.org/illegalimmigration-shiftingtide

Forslund, C. J. (1996). Analyzing pictorial messages across cultures. In D. C. Andrews (Ed.), *International dimensions of technical communication* (pp. 45–58). Arlington, VA: Society for Technical Communication.

Hall, E. T. (1976). *Beyond culture.* New York, NY: Anchor.

Hasnain-Wynia, R., Yonek, J., Pierce, D., Kang, R., & Hedges Greising, C. (2006). *Hospital language services for patients with limited English proficiency: Results from a national survey.* Chicago, IL: Health Research & Educational Trust and the National Health Law Program.

Hofstede, G. (2001). *Cultures and organizations: Software of the mind.* New York, NY: McGraw-Hill.

Holder, E. (2011). *Federal government's renewed commitment language access obligations under Executive Order 13166.* Retrieved July 10, 2013, from http://www.justice.gov/crt/lep/13166/AG_021711_EO_13166

Kreps, G. L., & Kunimoto, E. N. (1994). *Effective communication in multicultural health care settings.* Thousand Oaks, CA: Sage.

Languages Spoken at Home. (2009). *US Census*. Retrieved from http://factfinder.census.gov/servlet/STTable?_bm=y&-geo_id=01000US&-qr_name=ACS_2009_5YR_G00_S1601&-ds_name=ACS_2009_5YR_G00_

Limited English Proficiency (LEP). (2011). *Mission statement.* Retrieved from http://www.lep.gov/

Medical Translation. (n.d.). *Translation hub.* Retrieved from http://www.translationshub.com/types-translation/medical-translation.html

Miklavcic, P. R. (2004). *Migrant farm labor in Michigan: An analysis of recent trends in supply and demand, and policy implications.* Retrieved from http://www.aec.msu.edu/theses/fulltext/miklavcic_phd.pdf

New Mexico: Languages. (2010). *Worldmark encyclopedia of the states.* Retrieved from http://www.city-data.com/states/New-Mexico-Languages.html

Ogilvy Public Relations Worldwide. (2005). *Human papillomavirus creative materials testing target audience focus group research: Final report.* Gardosil Cervical Cancer Vaccine: Human papillomavirus essentials guide on CD-ROM. Progressive Management.

Perkins, J. (2003). *Ensuring linguistic access in health care settings: An overview of current legal rights and responsibilities.* Menlo Park, CA: Henry J. Kaiser Family Foundation.

Regulatory language requirements and the European Union. (2005). *Translate.com.* Retrieved from http://www.translate.com/language_tech_center/white_papers/regulatory_language_requirements_and_the_eu.aspx

Roberts, C. M. (2001). Meeting the needs of patients with limited English proficiency. *The Journal of Medical Practice Management, 17*, 71–75.

Sampson, A. (2006). *Language services resource guide for health care providers.* Washington, DC: National Health Law Program.

Schroeder, D. (2002). *Limited English proficiency (LEP) regulations: Implications and guidance for US hospitals serving minority populations.* Roswell, GA: DrTango.

Shah, A. (2008). Immigration. Global issues: Social, political, economic and environmental issues that affect us all. *Global Issues.* Retrieved from http://www.globalissues.org/article/537/immigration

St. Germaine, N. (2009). Localizing medical information for US Spanish-speakers: The CDC campaign to increase awareness about HPV. *Technical Communication, 56,* 235–247.

St. Germaine-Madison, N. (2007). *The effect of translation and localization changes on the efficacy of the American college of obstetricians and gynecologists' patient information pamphlets for a West Texas Tejano population.* PhD dissertation, Texas Tech University.

Tebeaux, E. (1999). Designing written business communication along the shifting cultural continuum. *Journal of Business and Technical Communication, 13,* 49–85.

U.S. Department of Health and Human Services. (2001). *Cultural competence works: Using cultural competence to improve the quality of health care for diverse populations and add value to managed care arrangements.* Washington, DC: Health Resources and Services Administration, DHHS.

Wong, C. (2004). Language assistance services in US health care institutions: Breaking through barriers. *Brandeis Graduate Journal, 2.* Retrieved from http://www.brandeis.edu/gsa/gradjournal/2004/wong2004.pdf

http://dx.doi.org/10.2190/LIIC7

CHAPTER 7

Usable DRM: Sailing the Pirate's Seas for Legal Entertainment

Liza Potts

Across the Internet, there are multiple locations for people to acquire entertainment content. Often, participants struggle through cumbersome processes in order to locate, watch, and interact with content. Their experiences are riddled with usability issues that create a deterrent for legal use. Such experiences lead many consumers to make choices between legal and illegal acquisition, watching advertisements in exchange for receiving free content, mistaking legal content for illegal, and surfing peer-to-peer (P2P) file-sharing systems for often illegal content. The purpose of this chapter is to trace these issues using methods drawn from Actor Network Theory (ANT) in order to illuminate the networks and usability concerns that participants confront as they look, view, or share content.

As technical communicators, we are best situated to contribute to this conversation because of our history of user advocacy and our involvement in the development process. Adopting and applying theories such as ANT provide us with methods for analyzing the various actors within communication networks, thus adding additional methods to our toolbox as both researchers and designers. In essence, ANT allows us to diagram the active people, organizations, activities, and technologies relevant to the user experience concerns surrounding key international issues. As a result, technical communicators can use an ANT approach to better pinpoint breakdowns in experience as a call for improving the user experience associated with important international interactions. Within the context of this chapter, we can use the ANT approach to understand the decisions the entertainment industry has made with regard to digital rights management (DRM) and the Digital Millennium Copyright Act (DMCA).

Examining these actor networks is of particular importance to technical communicators who are responsible for the information design of these systems, for these networks, and who influence the business rules that affect design decisions. By exploring and visualizing which actors are involved in a given experience, technical communicators can be better prepared to deliver the appropriate information to their audiences, and legal scholars can advise on potential cases concerning DRM and the DMCA.

Drawing upon our rich scholarship in communication design and content strategy, technical communicators can contribute to these conversations by becoming more involved in the work of delivering, positioning, and distributing digital entertainment. By pinpointing instances of information exchange, knowledge diffusion, and community building, we, as technical communicators, can jointly inform legal scholars and content designers about the social, political, and economic issues affecting these situations. We can then map these instances, locating the networks of technologies, organizations, and people involved in these communication exchanges.

Technical communicators must examine the connections between participants who seek content and the other people and technologies that surround participants and form network relationships with them. Adopting theories such as ANT can aid in this work. Doing so allows technical communicators to better understand the interactions, systems, and processes that are necessary for designing more usable experiences in general and in DRM systems specifically in the cases discussed in this chapter. The various participants who form such networks, along with the technologies they use, are intimately connected to the legal and policy constraints intended to govern the ways participants use content. Yet these connections are not always clear. This situation affects not only the usability of DRM systems but also the types of activities that many participants may perform in order to access entertainment content.

Because of the nature of our work, technical communicators are able to see across both these internal development processes and the resulting experiences of the users of these systems. As an extension of this work, we can influence the strategies of deployment and communication between these two parties in the digital entertainment field within global contexts. The approaches discussed in this chapter can aid technical communicators in addressing these complex legal issues by providing techniques and strategies that can contribute to the larger debate surrounding digital entertainment.

BACKGROUND ON DRM AND DMCA

As the creation and distribution of digital content has become increasingly easier, many copyright holders have turned to DRM technologies to control access to licensed content and to limit what participants might do with that content. The topic has already been the subject of many books (Boyle, 2008; Lessig, 2008; Litman, 2006; Logie, 2006; Vaidhyanathan, 2005) as well as journal articles (P. Allen, 2008; Arnab & Hutchinson, 2005; Erickson, 2003;

Potts, 2010; Samuelson, 2003). In 2003, *Communications of the Association of Computing Machinery* published a special issue devoted to DRM technologies. Further, as Samuelson (2003) notes, many organizations within the entertainment industries around the world advocate DRM technologies as solutions that "will prevent infringement on commercially valuable digital content" (p. 41).

Popular commercial software that use extensive DRM systems include Apple's iTunes (as well as their mobile devices, the iPod, iPhone, and iPad), one of the most widely used digital distribution services available on the web. A number of systems for accessing web-based content, such as game consoles and Blu-ray disc players, also use DRM to regulate what content or software those systems can run. In addition, many hard copy formats, including both DVD and Blu-ray discs, make use of DRM technologies. Common strategies for controlling such access include watermarking and fingerprinting content. According to Felten (2009), watermarking content occurs when a distinctive visual cue, piece of computer code, or sound is inserted into content so that it becomes "distinctively recognizable." Fingerprinting is very similar in principle but relies instead on "finding some distinctive feature of content that is already there" and is not easily removed.

These technologies are tied to a key question concerning the ownership of content that participants purchase. As Arnab and Hutchinson (2005) state, content containing DRM systems such as watermarks or fingerprints are "often sold as the user 'buying' the digital work," but legally, the "process is more like licensing" (p. 3). In other words, in international contexts, though someone may purchase a movie on a Blu-ray disc from a local retail store, that person might not legally own that content to the degree that he or she can do anything with it as he or she wishes.

This question was at the heart of the infamous Napster case in which the U.S.-based company set up a global P2P file-sharing system that allowed its participants to transfer content from one hard drive to another with no fees or purchases required. After numerous legal battles, the company was forced to close its doors and shut down its network. Many of its users were also targeted with multimillion dollar lawsuits claiming copyright infringement. In more recent cases, YouTube has become a major site of tension over the uploading and distribution of illegal content that is not officially licensed to YouTube or its participants. In fact, even legal streaming services such as Netflix and Hulu often provide content that is then downloaded illegally and distributed through other systems, including YouTube.

Interestingly, the laws regarding the use and scope of DRM vary widely from one country to the next. Attempts, however, have been made to create some degree of legal harmonization at the international level. The United Nations World Intellectual Property Organization (WIPO) advocated for an international set of regulations regarding the enforcement of copyright. These regulations, known as the WIPO Copyright Treaty (WCT), were signed in 1996 by many UN member nations, including the United States. In 1998, the United States passed the Digital Millennium Copyright Act (DMCA) in accordance with the WCT. The European Union passed the European Directive on Copyright 3 years later.

Unfortunately, not every nation is a member of the WCT, meaning that content hosted on servers in non–member nations is not subject to the same regulations as in member nations. Even within different WTC member nations, the laws regarding copyright and the use of DRM do not always transfer clearly from one country to the next. From a usability perspective, the key flaw in these strategies is that though they dramatically impact the usability of web-based content, none of them were ever constructed with that usability in mind. As this chapter will show, poor usability often traps participants in badly designed experiences that may inadvertently prompt illegal practices due to nothing more than the mounting confusion or frustration of participants.

The purpose of this chapter then is to situate an understanding of DRM technologies and international copyright issues as a key component to the work of technical communicators who are designing content management and content distribution systems within this context. Laws such as the DMCA directly affect what people are permitted do with web-based content. For technical communicators, such legal limitations also directly affect our work as experienced designers who build systems for participants on the web. To be a user advocate now means thoroughly understanding how these policies impact the experiences of those for whom we are advocating. We have a rich history of scholarship focused on theories, strategies, and techniques to improve technical communication and information design that are needed to enrich these conversations in business, computer science, and media studies. If we are to design improved experiences—through interfaces, documents, policies, and such—then this understanding of international legal limitations and user expectations is a crucial component of our future research.

MAPPING DIGITAL ENTERTAINMENT EXPERIENCES

Diagramming patterns, processes, and networks is a key procedure in the design cycle for information systems. Using lines to signal relationship strength, diagramming seeks to visualize what Thatcher (2001) calls a "generalized pattern" of connections among multiple cultures. As an issue that extends across international and cultural boundaries, the usability of DRM is embedded within such generalized patterns. Thus, diagramming the connections among people, groups, and technologies is a critical step that can uncover and highlight the major questions regarding intellectual property and cultural contexts that affect the usability of content managed by DRM systems. Tracing these connections allows us to understand not only the logic of such systems but also to find missing procedures and steps that might be helpful to participants (Potts, 2010).

The research and design team for DRM-restricted systems and content must understand the people, groups, technologies, events, and practices involved in these experiences before other design activities take place. Whether on paper or in digital visualizations, mapping these components sheds light on potential usability issues before such systems are implemented with immense investments in time and money. In the case of DRM technologies implemented in the wake of the DMCA and other similar policies, diagramming these experiences helps researchers understand the ways that DRM systems will affect other participant practices. In addition, such diagrams highlight the ways in which participants enlist other tools and technologies into their processes for finding and using digital content. And in international contexts, such mapping becomes essential to understanding when, where, and how various national and cultural factors might affect behaviors. No other major methods allow researchers and designers to map experiences in this way.

As I have argued elsewhere (Potts, 2010), ANT can supply this method. ANT traditionally argues for the *general symmetry* of agency among people and technology, meaning that both equally possess agency as "actors" embedded within social connections (Callon, 1986; Latour, 1996, 1999; Law, 1992; Mol & Law, 1994). Extending the actor concept to include organizations and processes important to these connections allows technical communicators to develop a fuller picture of the components that affect participant experiences within such networks (see Figure 1).

Though ANT insists on general symmetry of agency among actors, in practice, it is more realistic to assume that people have more agency than the tools and devices they use. Diagramming with ANT is not intended to view people, groups, and technologies as equal agents, but to view the connections among different actors as equally important to the participant experience. Other scholars have also found ANT useful for researching issues in technical communication related to the connections among people, groups, and technologies (McNely, 2009; Potts, 2008, 2009a, 2009b, 2010; Spinuzzi, 2008; Swarts, 2009).

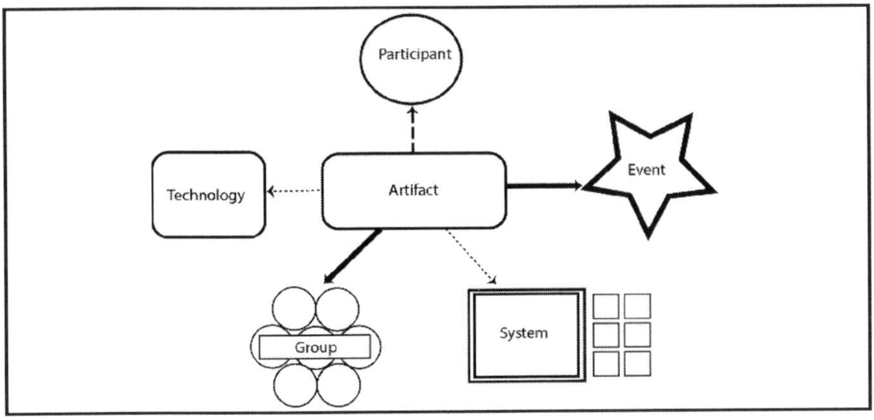

Figure 1. Example of an actor network diagram describing an assemblage of actors.

With respect to DRM and related international legal issues, identifying the actors and the linkages they have to one another within a network allows us to understand how participants find and consume digital content both within and in conflict with the legal regulations implemented through DRM. In doing so, we can assess the impact of DRM on the usability of digital systems that house, distribute, and play media content. Other diagramming methods, such as Unified Modeling Language (UML), focus on tracing specific interactions for accomplishing specific tasks within specific systems (Fowler, 2003). These methods focus on a single user interacting with a single piece of software. By contrast, ANT diagramming is intended to view these experiences in broader contexts, situating the participant experience within a larger ecology of people, groups, and technologies. Thus, the actors that affect international DRM-regulated experiences include many technologies, policies, and organizations that other methods, such as UML, often ignore (Potts, 2009a, 2009b). For systems that provide digital content on the web, especially for participants interacting through social web tools such as social networking systems, the usability of DRM technologies is tied to the linkages of participants and content to other actors.

This method, especially when combined with other approaches common in technical communication, provides a strong basis from which to explore the usability of content moving through DRM-regulated systems and networks. The implications of legal constraints and technologies that implement these constraints are extensive. We technical communicators must therefore begin to examine the ways such network actors influence the experiences of participants searching for and viewing content that falls under such regulation. ANT-based diagramming empowers researchers and designers to better understand contexts for participant activities, the relationships actors have to one another, and the ways participants pursue their activities.

The Case(s) for Usable Digital Entertainment

Digital content has become highly mobile. It is now capable of moving throughout the Internet and crossing national and cultural boundaries in the process. As a result, participants who use the web to find, watch, and play content are greeted by increasing concerns about the legality of their activities. While the participants themselves might be sitting quietly in front of their computers or using their smart phones while riding in a cab, the content they use often traverses these boundaries regularly. And this situation could potentially lead to usability problems as various legal and policy restrictions might affect participants' activities. Technical communicators must therefore be aware of the legal and technical issues that await online participants as they seek to not only watch or listen to digital media but to use it for their own purposes.

In his article entitled "Is the Sky Falling on the Content Industries?" (2010), legal scholar Mark A. Lemley traces myriad claims content owners made about how new technologies are causing the imminent destruction of their businesses. From gramophones to VCRs, Lemley discusses how notable figures such as John Phillip Sousa and Jack Valenti reacted to shifts in content distribution and usage. Across all of these claims, Lemley shows how these inventions actually led to increased revenue and product distribution rather than the destruction of these content-based industries.

Today we are faced with similar claims from industries that distribute music, television, movies, books, games, and other content. Representatives of these industries argue that the increased mobility of content, combined with the increased capacity of participants to adapt and repurpose that content, threaten these industry's ability to control the use of such media and to generate financial revenue from it. Organizations such as the Recording Industry Association of America point to cases like Napster as examples of those threats. Instead of adapting to the technological changes that do challenge traditional models of intellectual property and capital revenue, many of these industries have turned to tools such as DRM to extend those models into the era of digital distribution.

In doing so, the usability of entertainment media is pushed to the side or ignored, and in some instances, this situation might actually prove counterproductive to the goals of the content industries. In this way, DRM does not so much preserve copyright integrity for stakeholders as it "change[s] consumer expectations about what they are entitled to do with digital content" (Samuelson, 2003, p. 41). The usability of DRM is not concerned with adapting to the needs of participants. Rather, it attempts to redefine what those needs should be. Many of these attempts have stumbled or failed due to their poor usability. As Peter J. Allen (2008) states, DRM technologies have a "history of frustrating consumers with compatibility problems" that are combined with "unreasonable restrictions on how legally acquired digital media can be consumed" (p. 5). The EU has made note of these issues as well, acknowledging "potential obstacles to the free flow of information and additional burdens for economic operators and citizens" (p. 3).

As user advocates, technical communicators can use ANT to map the networks and interactions favored by participants. The visualizations provided by ANT-based diagramming better prepare technical communicators to present participant experiences to stakeholders and in turn impact the types of processes and regulations those stakeholders might favor (Potts, 2010). As Jo Allen (1990) points out, "No definition will adequately describe what we do" as technical communicators (p. 76). The increasing interdisciplinarity of technical communication leaves us well suited to tackle communication problems at the intersection of network design, entertainment, intercultural communication, and the law. The cases below demonstrate the issues relevant to technical communicators in these spaces, which have traditionally fallen outside the view the discipline. The usability of DRM within contexts that connect legal policies to technical systems and the use of content they mediate is certainly a core concern for technical communication, especially as similar concerns expand beyond the entertainment industry to other spaces, such as the classroom or workplace organizations.

The design of current implementations paradoxically provides robust tools for the quick and effective distribution of content while simultaneously funneling participants through processes that are not user-friendly. Discussing ANT, Latour (2005) argues, "The task of defining and ordering the social should be left to the actors themselves" (p. 23). To that end, participants who actually use DRM-regulated systems should be injected back into the design and testing processes for such systems, especially given the prospect that poor usability can actually defeat the very purposes for which DRM technologies are used.

Reading by Kindle Light

The Amazon Kindle is a thin, lightweight digital "reading device" that uses wireless connectivity to let owners download books, magazines, or newspapers and look at blogs posted online (Amazon, 2009). The device connects its owners to Amazon's wireless service for the Kindle, giving them access to over 350,000 pieces of content. Onboard tools also allow Kindle owners to annotate content using bookmarks, in-text highlighting, and clippings of text. In this section, I examine how the Kindle is used to change the model of textual ownership so that Kindle owners actually fall into the licensing model described by Arnab and Hutchinson (2005). This approach allows Amazon to access Kindles and remove content that owners have downloaded. This ability has led to concerns from both consumers and privacy advocates regarding the rights of Kindle owners and the freedom of information.

The Kindle's design is lightweight and utilizes Amazon's Whispernet technology. These two factors combined make the device highly portable for owners. Amazon has also made the device available in other countries, and it provides access to downloadable content around the world (Amazon, 2009). This factor is important as many parts of Europe, the Middle East, Africa, and Asia have faster access than other areas of the world, including the United States (Cellmaps, 2009).

In the summer of 2009, a major debate erupted when Amazon abruptly deleted copies of George Orwell's *1984* from the devices of every Kindle owner. The company claimed it did so because it discovered a third-party publisher did not own the licenses necessary to sell the text to Kindle customers. On the surface, the move appears to enforce the copyright protections of a text that was improperly sold by a publisher. However, Amazon's measures demonstrate that "DRM does not actually implement the fundamentals of copyright law, and is rather a mechanism for enforcing licensing and contract restrictions on digital data" (Arnab & Hutchinson, 2005, p. 1). In this case, Kindle owners had only purchased licenses that gave them access to specific content. Amazon, in turn, could refuse that access based on a series of contingencies, such as copyright violations. Amazon could access Kindles and delete *1984* without providing any prior notice to Kindle owners.

To better understand what happened, it is important to examine the network across which these events transpired. Figure 2 maps the relationships among actors at work in this event. The lines connecting the central actor to other people, organizations, and technologies indicate the strength of each relationship.

Though Amazon gave customers a refund, the company waited until after it deleted content from devices to notify Kindle owners of the problem. Kindle owners quickly turned to the Kindle Community forum maintained on Amazon's own website to voice their displeasure (Kindle Community, 2009). While some speculated the event was due to *1984*'s themes of government control of information and language and the loss of privacy (Parry, 2009), others took a more serious view. One commenter argued,

> If Amazon can delete books from your Kindle, presumably they can also replace those books with altered versions—to correct mistakes, delete defamatory remarks, bowdlerize naughty bits or even change an author's arguments if they or their corporate friends didn't agree with it. And you would never know. (Wetcoast, 2009)

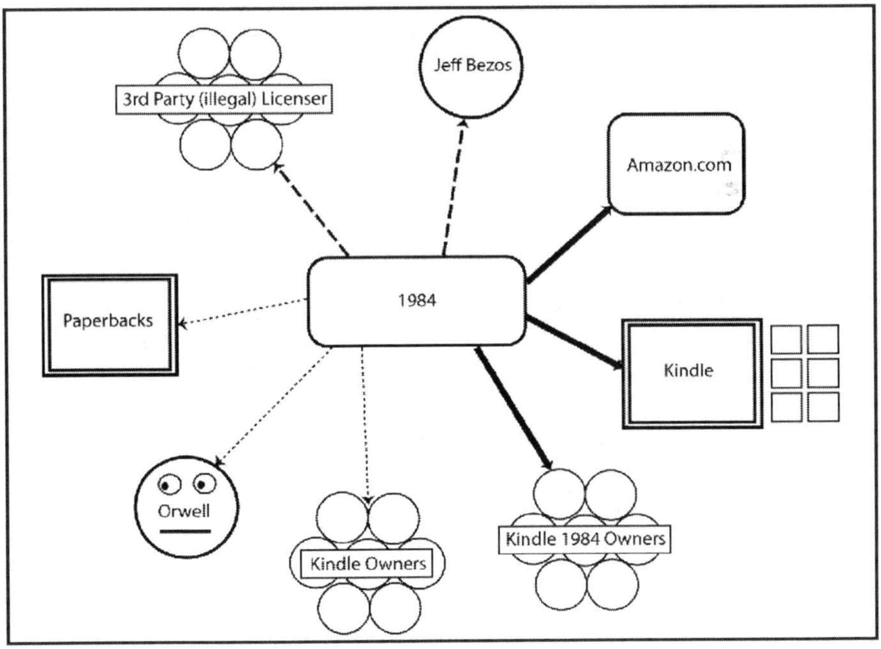

Figure 2. User experience of Kindle consumes (Potts, 2010).

Others implicitly hinted at the licensing models governing digital content, asking, "I wonder if Amazon will send representatives to customers' houses to retrieve dead tree copies?" (Caffeine Queen, 2009). Another individual made the comparison that "It's like having Barnes & Noble sell you a book, charge your Visa and then 3 months later change their mind, credit your card and DEMAND their book be returned" (Ron in Richmond, 2009).

These comments point to two issues deeply interconnected with the usability of DRM-regulated content. One is that while digital distribution has made information more easily accessible to millions of people, it has simultaneously made the regulation of distribution and copying of that content much easier, which also more restrictively defines what a customer's ownership of their purchased content is. As Arnab & Hutchinson (2005) note, the "enforcement of copyright law has always been passive in the physical world" (p. 1). In the digital world, connectivity gives enforcement a much more active and much further reach.

Second, the Kindle owners expressed immense confusion and frustration over this event, and DRM practices in general. One commenter, Sunny Lady (2009), expands the Barnes & Noble comparison:

> I liken it to a B&N clerk coming to my house when I'm not home, taking a book I bought from them from my bookshelf and leaving cash in its place. It's a violation of my property and this is the perfect example of why people (rightly) hate DRM.

Customers purchased a license for the content of the novel and not an actual physical copy of the novel itself. Amazon changes the model of ownership, giving the company more expansive rights to access Kindles and remove content. In Samuelson's (2003) terms, companies such as Amazon use DRM more to manage 'permissions' to do X, Y, or Z with digital information" (p. 42).

In this model, Kindle owners are simply purchasing permissions. Moreover, the Kindle's terms of service do not make clear to customers Amazon's right to access these devices and remove content. As a result, Kindle owners affected by this problem expressed outrage based on a sense that their copies were, in some sense, stolen from them without their permission. Yes, the company gave customers refunds, but those who had *1984* on their devices wanted the text and the right to read it at their leisure.

Clearly, the expectations of Kindle owners have not been balanced with the content-licensing practices of Amazon and its third-party partners. As Sunny Lady (2009) implicitly suggests, Amazon's business rules for selling properly licensed content are clearly disconnected from this specific case. Instead of balancing expectations with licensing practices, DRM gives Amazon a brute-force solution that ignores customer expectations all together. In the words of one commenter on a technology blog, "This is why DRM is bad, kids" (Anonymous, 2009).

Another commenter argued that the backdoor allowing Amazon to access Kindles and remove content should be closed (Frauenfelder, 2009). To put it another way, by violating the expectations of Kindle customers, Amazon helped create an atmosphere in which its practices are perceived as a direct attack on those customers. One reason for this is that DRM technologies that automate the regulation of content are not effective at adapting to situations in which policies may not so clearly apply (Erickson, 2003, p. 36).

The effects of this situation have already rippled throughout other countries as well. In France, book retailers have asked the French government to create and maintain an online hub through which texts can be sold and distributed electronically. Citing the Kindle incident as an example, they are seeking an alternative that mitigates these kinds of events (Laurent, 2010). Germany, in turn, implemented a similar system, dubbed Libreka. In such cases, there are certainly critical questions to be asked concerning national governments running such systems. These events, however, make it clear that many wish for an alternative to the digital retail and distribution systems that are currently in place. While U.S. citizens lament that, copyright extensions on American texts or films entered the public domain on the worldwide celebration of "Public Domain Day" (Masnick, 2011), residents of other countries without such restrictions were able to enjoy Trotsky, Fitzgerald, and Benjamin (Public Domain Works, 2011) on January 1, 2011 (Public Domain Day, 2011).

As a company that depends on solid relationships with its customers, Amazon arguably did not want to anger, frustrate, or alienate Kindle owners. However, it is also clear that the company's brute-force strategy, supported by DRM implemented into the Kindle experience, did anger, frustrate, and alienate many of its customers. The company ignored the perceptions of customer ownership and consumer rights to content they had purchased—an issue further complicated by a lack of clarity in the Kindle's terms of service regarding Amazon's ability to remove content. Mapping the Kindle owner's experience during the design process could have uncovered these issues. Moreover, in mapping this global case, technical communicators can highlight potential problems and become stronger user advocates who can influence both policy and implementation of agreements between companies and their customers.

More recently, Amazon has implemented a scheme that allows Kindle owners to lend the content they "purchased" for their Kindle. Figure 3 illustrates this approach and shows the relationship of a book (*The Cathedral and the Bazaar*), a publisher (O'Reilly Media), technologies (Kindle), content sellers (Amazon), and people (Lender, Borrower).

Essentially, books can be loaned to a Kindle owner for 14 days provided the text is "lendable"—a feature determined by the book publisher or rights holder (Amazon, 2013a). While the book is loaned out, the lender cannot read the book.

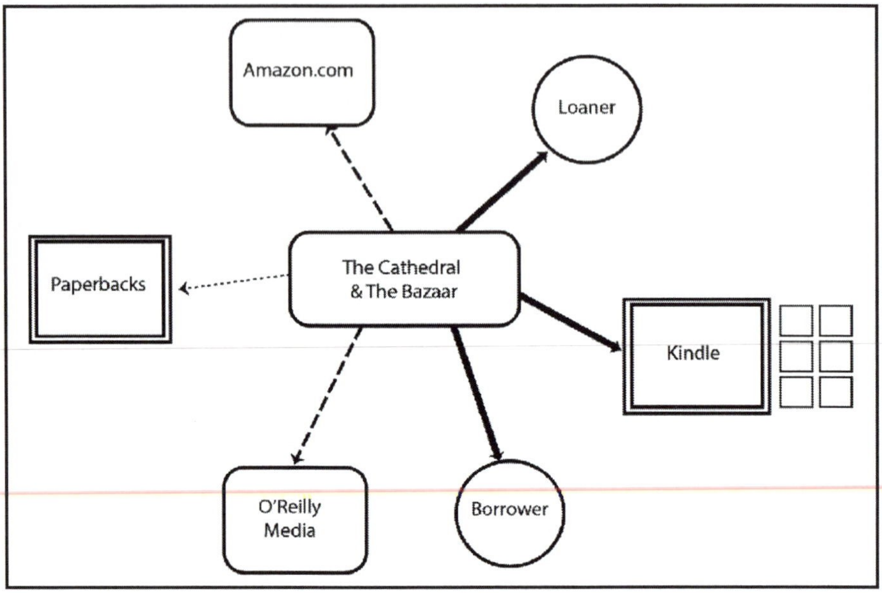

Figure 3. User experience of loaning Kindle ebooks.

Consumers can, in turn, find out if the book is available for exchange by looking up the book's product details on Amazon. If the *Product Details* list the book as "Lending: Enabled," the book can be loaned. The access to the book is then exchanged through Amazon's system, wherein the loaner must list the recipient's email address and their own name.

Announcing this new feature on Facebook, where it received 1,934 "Likes" and hundreds of comments, Amazon's status stated,

> The borrower does not need to own a Kindle. Kindle books can be read on Kindle or using our free Kindle reading applications for PC, Mac, iPad, iPhone, BlackBerry, and Android devices. The publisher or rights holder determines which titles are enabled for lending. (Amazon, 2011b)

Many of the comments listed under this status update were from consumers discussing their concerns for the new service. There were consumers lambasting Amazon for not allowing more books to be loaned, not extending the lending time, not allowing them to renew the loan, and other concerns. Many postings also discussed how other eBook services, such as the Nook, provide better lending services, including lending from local libraries. There were also calls for consumers to band together to tag books as "lendable" in order to encourage publishers to enable lending (Scottbooks, 2011).

With this lending scheme, Amazon is attempting to negotiate legal exchanges and discourage piracy. But how effective is this approach? A recent investigation undertaken by Magellan Media and book publisher O'Reilly set out to investigate whether or not pirating content led to fewer book sales: "Essentially, if the net impact of piracy is negative, then you would see sales fall off more quickly after piracy; if it were positive, the opposite" (Webb, 2011). In fact, the data for net sales of O'Reilly books showed an increase in sales for pirated material (O'Leary, 2009). However, the researchers temper these results by stating that there simply is not enough data to make generalizations to the rest of the publishing industry. They cite the other publishers as resistant to sharing their data for analysis, even when it is at "no-cost" to them (Webb, 2011). Clearly, researchers and designers need to further examine these issues to better prepare themselves for shaping user experiences.

Jumping Through Hulu Hoops

Television owners face similar issues. As television content has become exclusively digital in the United States, it is also common to access programming through a variety of methods. Televisions are no longer simply monitors, but can be directly linked via Internet connections to media streaming sites such as YouTube, Netflix, and Hulu. In addition, on-demand digital video services provide access to content, and other systems use digital video recording (DVR) to let owners record television broadcasts in digital format. In this increasingly connected, digitized space, television viewers can access virtually any content that can be housed in digital form, so long as that content is made available through a streaming or on-demand service. This section examines how the usability of these systems are deeply interconnected with the television viewers' habits for accessing and downloading content, whether they do so legally or illegally.

A 2009 Pew report suggests that 89% of Internet users between 18 and 29 years of age watch content on video-sharing sites (Madden, 2009). In many cases, these individuals do so because television networks and their related on-demand services do not always carry the content for which this audience is searching. Viewers may be searching for specific episodes that are not currently being broadcast by a television network. Thus, though DVRs and on-demand services provide many options for viewers to watch content at their own convenience, none of these services matter if the content is unavailable to begin with. As a result, many television viewers turn to the Internet for more options.

The Internet connects its participants to a wide array of options that access television content both illegally and legally. Arnab and Hutchinson (2005) note that "with the advent of load sharing peer-to-peer networks like BitTorrent, it is now virtually impossible to stop distribution of digital data on the Internet or to isolate all the sources of reproduction" (p. 1). As a result, sites like YouTube and BitTorrent P2P file-sharing systems provide ready illegal access to such content. They are accompanied by sites such as Pirate Bay, which serves to index content across the Web, as well as many other sites that allow Internet users to view content illegally. And these sites are growing exponentially, both in terms of number and usage.

Legal systems, such as Hulu, function similarly to Amazon, for they provide "as much content as possible [while] keeping it all on our service permanently." That content is regulated according to "streaming clearances" or other licenses "granted by our content partners" (Hulu, 2010). Television content on Hulu regularly becomes unavailable after a period of time and for a variety of reasons:

> An episode or movie may expire due to myriad reasons, including music clearances, impending DVD sales and syndication deals, among others. It's disappointing to us when a video must come down. As videos are the heart of our business, we're always working to continue to expand streaming clearances. (Hulu, 2010)

Hulu certainly works well to promote television programming and provide access to content viewers might otherwise have missed. As one Hulu commenter stated, "I missed *Buffy* in its original run, but I am now loving the chance to catch up. More seasons please!" (Macabri, 2009). Macabri is using Hulu to enjoy a television show that was originally broadcast from the late 1990s through the early 2000s. Her request for more seasons, however, hints at the potential usability problem stemming from content regulations and "clearances." Note that Hulu's statement does not indicate how, when, or why content may be pulled from the site. And not only does the removal of content potentially frustrate Hulu users, the lack of clarity about these issues can exasperate the problem as well.

Figure 4 illustrates the actor relationships for Hulu users trying to view *Buffy the Vampire Slayer* in October 2009. (Hulu's clearances for the show expired, resulting in the show being pulled from the site.)

In this figure, solid lines indicate the strongest relationships among actors (in this case the show itself), the audience viewing it on the Hulu site, and Hulu itself. One Hulu user, in turn, took to the discussion boards on Hulu to state both his love of the show and to plead for more seasons of the program:

> One of the best shows out there!!!! When can we get more??? I was almost done season three!!!! and then they took it off! wonderful acting and actors and of course vampires = one heck of a show! Please put on more seasons including three!!!! Im desperate to see more! (Pierson, 2009)

With the release of Hulu Plus (Hulu's pay system), there are now numerous options for fans seeking out television and movie content. However, even though these consumers are paying for this access, advertisements are still shown. It is unclear if both paying for content and having to watch advertisements will be welcome by consumers. Currently, those are the available models for cable

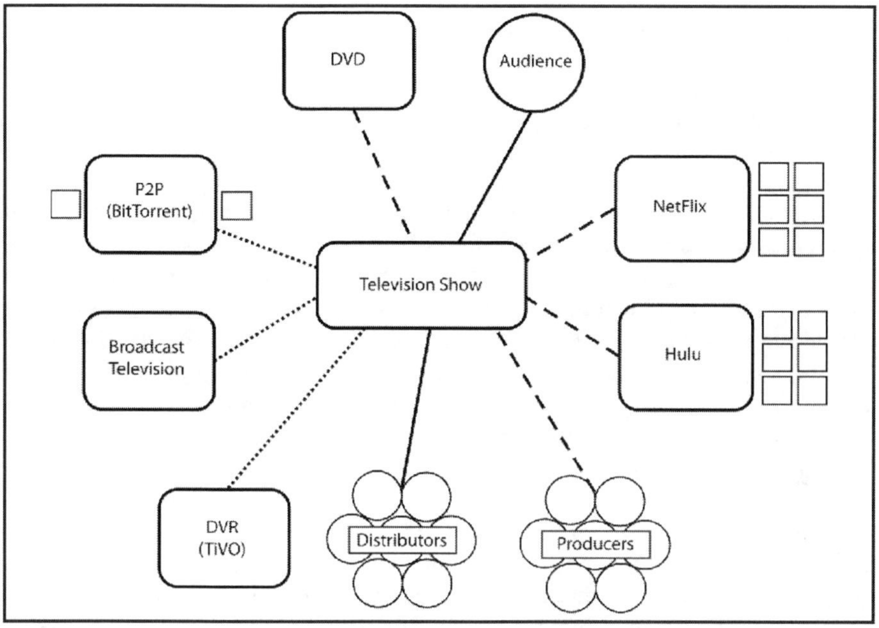

Figure 4. Viewing media content online (Potts, 2010).

television. The video provider Netflix, however, offers a premium cable model wherein no advertisements are shown. As more data flows in, we will have stronger answers to these questions.

By mapping the viewing experience with ANT, technical communicators could have uncovered the broader experience of using Hulu to watch television content. This process would have pointed out that Hulu visitors want a seamless experience with this content, unhindered by legal and policy issues that fragment those experiences. In this case, the policy of limited clearances creates a usability problem that leaves viewers with little option but to seek content elsewhere, perhaps illegally. Using ANT, technical communicators have the capacity to reveal these expectations and shape the policies that impact the viewing experience of television content through sites such as Hulu. In this case, the business rules that define Hulu's service do not align with its visitor's needs. Technical communicators can highlight this discrepancy and advocate for a more seamless integration of business rules with those needs and expectations.

Of Context and Country

Many of the issues discussed in this chapter point to the need for understanding these experiences from the perspective of the individual rather than the corporation or government. Internet participants turn to a number of different options for listening to music online. One of the most popular is Pandora, which bills itself as a "music discovery service" (Pandora, 2010a). Pandora presents an interesting case for examining the role that legal policies and DRM technologies play in defining what content or service is available in different countries and regions.

Pandora allows visitors to stream music directly from its site by using their preferences to create stations for certain types of music or artists. Using the "Music Genome Project," Pandora then injects these stations with music that is in some way related to the preferences defined by visitors (Pandora, 2010a). Visitors can further refine their stations by voting songs up and down, or by "seeding" stations with further song and artist choices. Customers who pay for subscriptions to the service can download a desktop application called Pandora One, which accesses their stations. Subscribing customers can also bypass the advertisements that are otherwise broadcast in visitors' stations.

Whether paying subscribers or simply as visitors, Pandora users must register with the site and provide a U.S. zip code that indicates their residency in the United States. (Due to various legal restrictions regarding music, Pandora is not available outside of the U.S.) The zip codes provided by registered users are cross-referenced with the geolocation of the Internet Protocol (IP) address assigned to the computer that is being registered. When a registered user returns to Pandora's site or opens the Pandora One application, their login information is again cross-referenced with the IP address of the computer they are using. This is done to determine whether or not the user can access Pandora content at that time.

Under this system, IP addresses in countries other than the United States are blocked. Anyone who attempts to login to the site from a location outside of the United States—including American citizens who have paid for subscriptions—is denied access. In this case, copyright laws are based on geographic location rather than on citizenship. Thus, the issue of usability in this case is not so much one of content mobility but one of user mobility. As a result, an American citizen traveling in a foreign country can be refused access to this service due to that person's geographic location, despite the individual having proven his or her citizenship according to Pandora's business rules.

Yet Pandora users and customers have not simply accepted this dilemma. Instead, they turn to various "spoofing" methods that allow them to trick Pandora into thinking they are in the United States when they might really be in another country. Spoofing is a common online strategy, and a Google search will immediately return many resources providing explanations and other support for spoofing servers. In its most basic form, spoofing uses a proxy server to give the user's computer a fake IP address that appears to be in the United States. This approach hides the computer's true identity and allows the visitor to circumvent Pandora's policies and DRM strategy. As a result, Internet participants have devised numerous spoofing methods (IBM, 2008). And while spoofing typically requires a level of technical skill and expertise beyond the range of most Internet participants, the availability of sites explaining such strategies widens the number of people who can use the strategy so long as they are proficient at using the Internet and computers. Such spoofing is a viable option for many of these issues discussed in this chapter, albeit an illegal one. Clearly, we need to work to provide experiences that consumers want rather than leaving the door open for illegal choices.

Figure 5 shows the ways Pandora visitors access content in the United States and in Portugal. Using a proxy server to spoof Pandora's servers, visitors in Portugal appear as U.S.-based users who can then listen to music without any copyright restrictions stemming from geographic location.

The most important feature of Figure 5 is the lack of a human actor. Instead, the diagram depicts relationships between countries and technologies. In so doing, it highlights the emphasis on geolocation rather than on individual citizenship. Where a person is located is more important in this experience than who that person may in fact be or whether that person has paid for the site's content. Interestingly, Pandora's frequently asked questions (FAQ) state this explicitly: "Only the current location of the listener matters for international licensing law—not citizenship, national origin, or country of permanent residence" or "source of revenue" (Pandora, 2010c). Pandora acknowledges this dilemma and states that they do have "the ultimate goal of offering our service globally" (Pandora, 2010b). Yet the legal complexities of copyright laws across international boundaries currently prohibit these efforts, complicating the usability of these systems and, again, forcing people to search elsewhere for content.

Lessig (2002, 2008) has argued for the need to balance the concerns of copyright stakeholders with the need for technological innovation. With respect

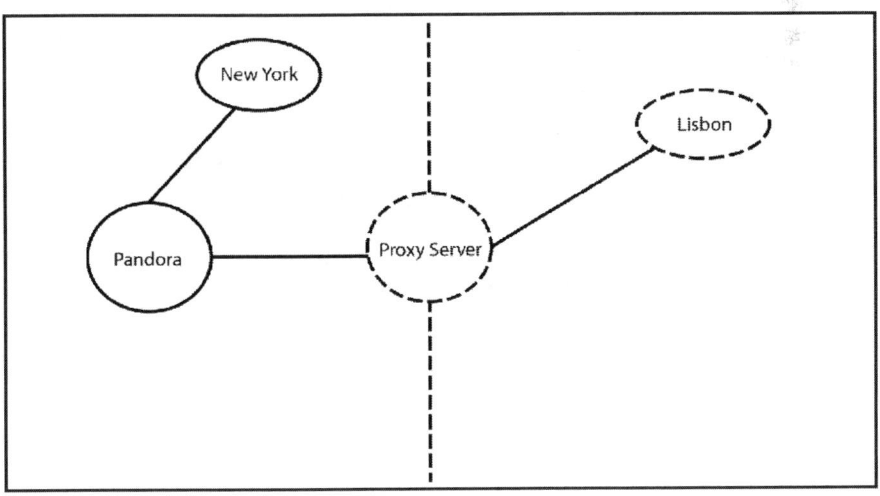

Figure 5. Accessing content across geographic boundaries (Potts, 2010).

to content accessed within the web, innovation includes the need to define use and participation in ways that better reflect the actual practices of online participants, such as those who access Pandora. In other words, for visitors who have provided the requisite proof of citizenship to Pandora—and especially if they have paid for subscriptions to Pandora's content—the notion that their citizenship does not matter runs counter to their expectations. As these users travel, they expect that they should be able to access content while on the go. That their geographical location stands in the way becomes a usability issue that is not only frustrating but also potentially encourages them to use other means to access this content.

THE EXPERIENCE OF US AND THEM

In the examples discussed in this chapter, the central theme of consumer experience is examined as one in which everyday people attempt to view, own, borrow, or distribute digital entertainment in global contexts. Overwhelmingly, content distributors such as Amazon, Hulu, Netflix, and Pandora work to make transparent the various policies of the rights holders that release this content and the governments who regulate them. In doing so, they are making visible the various networks that are involved with the economic exchange and legal restrictions involved in these experiences. Clearly, they also want to ensure that the consumer knows that it is not the content distributors' fault that they cannot serve up the content for the durations and methods that consumers may want. While it is noble of them to create such transparencies, it is questionable as to who is responsible for these disrupted experiences. And while various content acquisition "workarounds" exist, not all of them are appealing, and many of them are illegal.

The time spent exchanging DVDs through Netflix, for example, can be costly to consumers. Because many television programs are not digitized on Netflix, consumers must receive this content through physical DVDs. The experience of renting these DVDs is far from optimal. For many shows, only a handful of episodes are burned on each individual disc, causing a headache for the consumer who has to rent these DVDs, wait for the DVD to arrive, watch two episodes, return the DVD, wait for the next one in the sequence, and repeat this process. This experience of exchange, view, synchronize, and exchange is tiresome at best and incredibly frustrating at worse. In questioning these experiences, Gillespie (2006) states, "What is really at issue here is not just expertise and innovation, but a broader question of user agency: a software design community that encourages it, a regulatory constraint that forbids it" (p. 661). Netflix is attempting to move toward a digitally focused business, although it is unclear if the content industries will support this transition.

Such aggravation can make solutions such as illegal websites and P2P networks seem appealing to consumers who ordinarily would pay for content. In this light, we can argue that such systems support better user experiences because they deliver the content viewers want in delivery mechanisms they find more appealing. Lessig (2008) recognized this issue when he stated, "There exists not just the commercial economy, which meters access on the simple metric of price, but also a sharing economy where access to culture is regulated not by price, but by a complex set of social relations" (p. 145). We need to work toward delivering content to viewers in formats, streams, and contexts in which they want to receive them.

With regard to the Orwell incident, this issue becomes even more relevant. On Boing-Boing, a respected technology blog, one post noted that such moves by Amazon "will encourage readers to visit Websites in countries where the copyright has expired on Orwell's books so they can get free un-stealable electronic copies" (Frauenfelder, 2009). Such statements are clearly not in Amazon's favor; the DRM implementation on the Kindle is broken in this respect. Amazon's goal is to be the company "where people can find and discover virtually anything they want to buy online" (Amazon, 2013b). In order to encourage user experiences that can match Bezos' concept of Amazon's mission, the stakeholders parties (both corporate and governmental) will need to continue to look to improvements that can balance their business needs with consumer expectations.

Another issue is the complete lack of international experience in lending Kindle books. Amazon blames this issue on the publishers as well as geographical location, stating, "If a loan is initiated to a customer outside the United States, the borrower may not be able to accept the loan if the title is not available in their country due to publisher geographical rights" (Amazon, 2011a). This example draws us back to the issues experienced by Pandora users and again asks us to rethink these experiences from the perspective of the consumers rather than geographic location.

Emerging research in technical communication is working toward redefining our assumptions concerning international and intercultural issues aimed at investigations into "audience, purpose, rhetorical patterns, and document design" (Hayhoe, 2006, p. 141). Over the past decade, various studies have examined issues of ethics and intercultural issues (Voss & Flammia, 2007), design and localization (St.Amant, 2005), internationalization (Zahedi, Van Pelt, & Song, 2001), and cultural context (Wang & Wang, 2009). Giving insight into the difficulties of navigating cultural differences in writing and design are important for understanding how to build experiences for our users, audiences, and participants. With regard to laws, policies, and interfaces for technology, such an understanding of international issues is critical.

Across all of these examples, the networks of entertainment content actors are problematized. The wants of the consumers and the restrictions implemented by the content industries are in direct conflict. Through discourse on various sites, consumers are able to voice opinions and share their outrage. As we move toward more and more entertainment content being digitized over the next several years, these conflicts will only increase. It is up to us as advocates for the user experience to work on the design of systems that are more usable and the design of policies that are more user-friendly while still creating viable business models that can support these experiences.

CONCLUSION

St.Amant (2002) has called for a greater understanding of intercultural communication practices in computer-mediated communication environments. The cases in this chapter point to clashes among social, legal, business, and technical cultures that potentially frustrate the user experiences of those seeking media content on the web. The implementation of digital distribution systems and the DRM technologics that regulate their use are not user-centered systems that account for user experiences. Instead, they attempt to define what those experiences should be by regulating customer and user practices. They also assert far-reaching business rules that further limit customer ownership over their experiences. This is a call for technical communication scholars and practitioners to become more involved in these spaces as policies are updated, technologies are designed, and experiences are examined.

ANT provides a method with which technical communicators can begin to resolve some of these issues. With ANT, researchers and designers are able to see what users need and want within these current sociotechnical settings. We must design for users experiences of entertainment content that foster participation and usable experiences. By working to restrict access to and control of the use of content, current systems might in fact defeat their own regulations and policies by inadvertently encouraging users to turn to other options. ANT allows us to look across these ecosystems. It allows us to find and define the relevant people, organizations, groups, and technologies that make up the whole of the user experience. The knowledge gained from such applications can help technical communicators better situate the legal and policy frameworks currently used by different countries within the expectations of Internet participants who want more fulfilling experiences.

Currently, confusion and frustration are typical of these experiences, especially those regulated by DRM technologies and practices. By adopting theories such as ANT, technical communication scholars can add rich tools to our toolbox. In so doing, they can also add to our rich history of user advocacy and communication design. Tracing the flow of content, information, and even potential users through these systems allows us to design systems that can support stakeholders. It does so by taking into account the user experience rather than seeking to work around that experience or to simply squash it entirely. And such an approach is essential to success in the increasingly globalized context of the modern world.

REFERENCES

Allen, J. (1990). The case against defining technical writing. *Journal of Business and Technical Communication, 4,* 68–77.

Allen, P. (2008, September 1). Rip, mix, burn . . . sue . . . *ad infinitum*: The effects of deterrence vs. voluntary cooperation on non-commercial online copyright infringing behaviour. *first monday, 13*(9). Retrieved from http://firstmonday.org/htbin/cgiwrap/bin/ojs/index.php/fm/article/view/2073/2025

Amazon. (2009). *Kindle product page*. Retrieved from http://www.amazon.com/Kindle-Wireless-Reading-Display-Generation

Amazon. (2011). *Announcement: Kindle Book Lending now available*. Retrieved from http://www.amazon.com/forum/kindle/Tx2C5X2K2KCS71H?_encoding=UTF8&cdForum=Fx1D7SY3BVSESG&displayType=tagsDetail&ref_=cm_cd_cg_ef_tft_tp

Amazon. (2013). *Borrow books from the Kindle Owners' Lending Library*. Retrieved from http://www.amazon.com/gp/help/customer/display.html?nodeld=200757120

Amazon. (2013). *Overview*. Retrieved from http://phx.corporate-ir.net/phoenix.zhtml?c=176060&p=irol-mediaKit

Anonymous. (2009). Comment. *Boing-Boing*. Retrieved from http://boingboing.net/2009/07/17/amazon-zaps-purchase.html#comment-544611

Arnab, A., & Hutchinson, A. (2005). Fairer usage contracts for DRM. *DRM '05*. Alexandria, VA: Association for Computing Machinery.

Boyle, J. (2008). *The public domain: Enclosing the commons of the mind*. New Haven, CT: Yale University Press.

Caffeine Queen. (2009). Comment. *Amazon's Kindle community*. Retrieved from http://www.amazon.com/tag/kindle/forum

Callon, M. (1986). Some elements of a sociology of translation: Domestication of the scallops and the fishermen of St Brieuc Bay. In J. Law (Ed.), *Power, action, & belief. A new sociology of knowledge?* (pp. 196–229). London, UK: Routledge.

Cellmaps. (2009). *Kindle with U.S. and international wireless*. Retrieved from http://client0.cellmaps.com/tabs.html#cellmaps_intl_tab

Erickson, J. (2003). Fair use, DRM, and trusted computing. *Communications of the ACM, 46*, 34–39.

Felten, E. (2009). Everyone hates DRM. *The Intellectual Property Colloquium*. Retrieved from http://ipcolloquium.com/mobile/2009/04/drm/?type=CLE&id=34

Fowler, M. (2003). *UML distilled: A brief guide to the standard object modeling language* (3rd ed.). Reading, MA: Addison-Wesley Professional.

Frauenfelder, M. (2009). Amazon zaps purchased copies of Orwell's *1984* and *Animal Farm* from Kindles. *Boing-Boing*. Retrieved from http://boingboing.net/2009/07/17/amazon-zaps-purchase.html

Gillespie, T. (2006). Designed to "effectively frustrate": Copyright, technology and the agency of users. *New Media & Society, 7*, 651–669.

Hayhoe, G. (2006). Needed research in global technical communication. *Technical Communication, 53*, 141–142.

Hulu. (2010). Expiration. *Programming Info*. Retrieved from http://www.hulu.com/support/content_faq

IBM. (2008). Spoofing. *IBM Internet security system*. Retrieved from http://www.iss.net/security_center/advice/Underground/Hacking/Methods/Technical/Spoofing/default.htm

Kindle Community. (2009). Mysterious George Orwell refunds. *Amazon*. Retrieved from http://www.amazon.com/tag/kindle/forum

Latour, B. (1996). *Aramis. Or the love of technology* (C. Porter, Trans.). Cambridge, MA: Harvard University Press. (Original work published 1993)

Latour, B. (2005). *Reassembling the social*. New York, NY: Oxford University Press.

Laurent, L. (2010). French retailers urge national e-book hub. *Reuters*. Retrieved from http://www.reuters.com/article/idUSTRE60C4EO20100113

Law, J. (1992). Notes on the theory of the actor-network: Ordering, strategy and heterogeneity. *Systems Practice, 5,* 379–393.

Lemley, M. A. (2010, August 10). Is the sky falling on the content industries? *Social Science Research Network.* Retrieved from http://papers.ssrn.com/sol3/papers.cfm?abstract_id=1656485

Lessig, L. (2002). *The future of ideas*. New York, NY: Vintage.

Lessig, L. (2008). *Remix: Making art and commerce thrive in the hybrid economy*. New York, NY: Penguin.

Litman, J. (2006). *Digital copyright*. Amherst, NY: Prometheus.

Logie, J. (2006). *Peers, pirates, and persuasion: Rhetoric in the peer-to-peer debates*. West Lafayette, IN: Parlor.

Macabri. (2008). Comment. *Hulu*. Retrieved from http://www.hulu.com/buffy-the-vampire-slayer

Madden, M. (2009). The audience for online video-sharing sites shoots up. *Pew Internet & American Life Project*. Retrieved from http://www.pewInternet.org/Reports/2009/13—The-Audience-for-Online-VideoSharing-Sites-Shoots-Up.aspx

Masnick, M. (2011, January 3). US is left waiting for Godot on Public Domain Day: Once again, absolutely nothing enters the public domain this year. *TechDirt*. Retrieved from http://www.techdirt.com/articles/20110102/13551712487/us-is-left-waiting-godot-public-domain-day-once-again-absolutely.shtml

McNely, B. (2009). Backchannel persistence and collaborative meaning-making. *Proceedings of the 27th ACM international conference on Design of Communication*. Bloomington, IN: SIGDOC.

Mol, A., & Law, J. (1994). Regions, networks and fluids: Anaemia and social topology. *Social Studies of Science, 24,* 641–671.

O'Leary, B. (2009). *Impact of P2P and free distribution on book sales*. Sebastopol, CA: O'Reilly Media.

Pandora. (2010a). What is Pandora? *Pandora*. Retrieved from http://blog.pandora.com/faq/contents/13.html

Pandora. (2010b). Can I listen to Pandora outside the United States? *Pandora*. Retrieved from http://blog.pandora.com/faq/contents/79.html

Pandora. (2010c). Additional Pandora international Q&A. *Pandora*. Retrieved from http://blog.pandora.com/faq/contents/446.html

Parry, D. (2009). Comment. *Amazon's Kindle Community*. Retrieved from http://www.amazon.com/tag/kindle/forum

Pierson, A. (2009). Comment. *Hulu*. Retrieved from http://www.hulu.com/buffy-the-vampire-slayer

Potts, L. (2008). Designing with actor network theory: A new method for modeling holistic experience. *Proceedings of the International Professional Communication Conference*. Piscataway, NJ: IEEE.

Potts, L. (2009a). Designing for disaster: Social software use in times of crisis. *International Journal of Sociotechnology and Knowledge Development, 1*(2), 33–46.

Potts, L. (2009b). Using actor network theory to trace and improve multimodal communication design. *Technical Communication Quarterly, 18,* 281–301.

Potts, L. (2010). Consuming digital rights: Mapping the artifacts of entertainment. *Technical Communication, 57,* 300–318.

Public Domain Day. (2011). 1 January 2011. *Public Domain Day*. Retrieved from http://publicdomainday.org/

Public Domain Works. (2011). Persons whose works enter the public domain in 2011. *Public Domain Works*. Retrieved from http://www.publicdomainworks.net/stats/year/2011

Ron in Richmond. (2009). Comment. *Amazon's Kindle Community*. Retrieved from http://www.amazon.com/tag/kindle/forum

Samuelson, P. (2003). DRM {and, or, vs.} the law. *Communications of the ACM, 46,* 41–45.

Scottbooks. (2011). *Hello fellow Kindlers*. Retrieved from http://www.amazon.com./forum/kindle%20lending%20enabled/ref=cm_cd_et_md_pl?_encoding=UTF8&cdForum=Fx311IUDYPPMK2C&cdMsgID=Mx279JWYGCP36MC&cdMsgNo=1&cdPage=1&cdSort=oldest&cdThread=Tx11AV5F13468HP#Mx279JWYGCP36MC

Spinuzzi, C. (2008). *Network: Theorizing knowledge work in telecommunications*. New York, NY: Cambridge University Press.

St.Amant, K. (2002). When cultures and computers collide: Rethinking computer-mediated communication according to international and intercultural communication expectations. *Journal of Business and Technical Communication, 16,* 196–214.

St.Amant, K. (2005). A prototype theory approach to website localization: An analytical method for technical communicators. *Proceedings of the International Professional Communication Conference*. Piscataway, NJ: IEEE.

Sunny Lady. (2009). Comment. *Amazon's Kindle Community*. Retrieved from http://www.amazon.com/tag/kindle/forum

Swarts, J. (2009). The collaborative construction of "fact" on Wikipedia. *Proceedings of the 27th ACM international conference on Design of Communication*. Bloomington, IN: SIGDOC.

Thatcher, B. (2001). Issues of validity in intercultural professional communication research. *Journal of Business and Technical Communication, 15,* 458–489.

Vaidhyanathan, S. (2005). *The anarchist in the library: How the clash between freedom and control is hacking the real world and crashing the system*. New York, NY: Basic.

Van Eijk, N., Poort, J., & Rutten, P. (2010). Legal, economic and cultural aspects of file sharing. *Communications & Strategies, 77,* 35–54.

Voss, D., & Flammia, M. (2007). Ethical and intercultural challenges for technical communicators and managers in a shrinking global marketplace. *Technical Communication, 54,* 72–97.

Wang, Y., & Wang, D. (2009). Cultural contexts in technical communication: A study of Chinese and German automobile literature. *Technical Communication, 56,* 39–50.

Webb, J. (2011). Book piracy: Less DRM, more data. *O'Reilly TOC*. Retrieved from http://toc.oreilly.com/2011/01/book-piracy-drm-data.html

Wetcoast. (2009). Comment. *Boing-Boing*. Retrieved from http://boingboing.net/2009/07/17/amazon-zaps-purchase.html#comment-544653

Zahedi, F., Van Pelt, W. V., & Song, J. (2001). A conceptual framework for international web design. *International Transactions on Professional Communication, 44*(2), 83–103.

PART III

Ownership and Authorship

http://dx.doi.org/10.2190/LIIC8

CHAPTER 8

Workplace Realities versus Romantic Views of Authorship: Intellectual Property Issues, Globalization, and Technical Communication

Pavel Zemliansky and Traci A. Zimmerman

The environments in which professional writers work are changing rapidly. The seemingly constant evolution of virtual writing and collaboration technologies and the globalization of the workplace are two of the most important factors that effect this change. These and other developments have prompted a reexamination of the legal aspects of professional writers' work. No doubt, a large part of this reexamination has to do with copyright laws and regulations, which are tightly connected to the questions about text and information ownership and the rights to share. In his March/April 2010 article in *EDUCAUSE Review*, Lawrence Lessig addresses this concern as the "extraordinarily large elephant in the room": the fact that copyright has grown from having a "tiny" role in our lives to being the gate through which we have to pass in order to have access to our culture. Why has "the architecture of digital technology" become so inextricably tied to "the architecture of copyright law" (p. 28)?

In order to answer this question meaningfully and usefully for professional writers and professional communication students, we members of the profession need to consider some even more fundamental questions and problems. Has, for example, the very nature of "text" been changed by new computer-mediated communication technologies and processes? What about "the author" or "the audience"? How much premium are we as educators willing to put on our students' ability to produce individual work as opposed to their ability to collaborate with other writers, including writers in other cultures or economic systems?

Answering, or at least considering these questions will allow us as members of the profession and as educators to design educational philosophies, methodologies, courses, and assignments fitting the globalized nature of professional communication in the 21st century. As professional writing and communication changes, so should our understandings of authorship and textual ownership, intellectual property, and our rights and responsibilities as professional communicators.

In this chapter, we argue that educational practice in training future practitioners in the field of professional writing and communication has not always kept up with professional practices when it comes to authorship, text ownership, copyright, and collaboration. By tracing some of the histories, theories, and questions that form our understanding of the field and of what professional writers do, we will identify how the questions of the past can help us address the challenges of the present and what implications they have to teaching, learning, and future research. Our ultimate goal is to highlight areas wherein theory and practice disconnect and to begin to chart a path toward realignment of professional practice and training of future professional communicators for the global marketplace.

This change, however, should not be a one-way movement, with educators being the only ones who make changes in their work to "adjust" to professional practices. While such adjustments might be necessary for immediate practical purposes (e.g., graduates getting jobs), we believe that both academy and industry have much to learn from one another. Therefore, we hope that our work will be relevant both to educators and to industry practitioners. In order to bring the two sides closer together, we highlight some key issues of concern both to academics and to working professionals. We then propose some approaches and strategies for bringing pedagogy and praxis closer together.

INITIAL CONTEXT

In his 1996 article "Who 'Owns' Electronic Texts," Tharon Howard presaged the global and legal issues that technical communicators would face in the coming century. According to Howard, "Recent trends toward more collaborative writing projects in the workplace . . . are all challenging the popular, romantic view that an author owns his or her own text" (p. 251). Furthermore, Howard states, "More and more frequently, professional writers are finding themselves confronted with intellectual property and copyright issues [they] are . . . unprepared to deal with" (p. 251). Though it has been almost two decades since Howard identified these issues, we have come no closer to understanding them. If anything, the landscape has become even more complicated.

Since Howard (1996) made his observations, Web 2.0, "social" web design, the Creative Commons movement, the single-sourcing of workplace documentation, globalization, and other developments have made the learning and teaching of legal and intellectual property issues even more urgent for technical communicators. As Jessica Litman (1994) writes,

> Technology, heedless of law, has developed modes that insert multiple acts of reproduction and transmission—potentially actionable events under the copyright statute—into commonplace daily transactions. Most of us can no longer spend even an hour without colliding with the copyright law. (pp. 34–35)

Downloading, forwarding, posting, even just using the Internet produces "copies," "thus triggers copyright" (Lessig, 2010, p. 26).

But this "collision" with copyright affected our lives long before now. Most 20th century inventions, like the Xerox copier, allowed us to "manage important originals without destroying them, [make] copies almost as good as the originals, and [allow] all this wizardry [to be] within reach of everybody" (Wirten, 2004, p. 63). And like its 21st century counterparts, the copier faced its share of critics. In fact, media critic Marshal McLuhan called "xerography" the "most startling and upsetting electric innovation" that would bring "a reign of terror into the world of publishing" (as cited in Wirten, 2004, p. 66). His primary point of resistance?

> Every reader can become both author and publisher . . . [that] anyone can take a book apart, insert parts of other books . . . and make his own book in a relatively fast time . . . xerography is electricity invading the world of typography, and it means a total revolution in this old sphere. (Wirten, 2004, p. 66)

The "revolution" is ongoing, but "electricity" is no longer "invading" the world of typography. Rather, it has begun to transform it, interrogate it, and remix it. In effect, the technologies that we use to access the world are constantly changing, as are the laws that attempt to regulate this access. But our understanding of authorship has tended to remain rather static despite the unprecedented access and opportunity that authors have in the digital age. As a result, many students of technical communication hold a "romantic" view of the author and of their texts as solely the student's own intellectual property, a sentiment reinforced by years of academic practices that privilege "solo-authored" work.

Only relatively recently has technical communication even been considered an "authored" form of discourse. Slack, Miller, and Doak (1993) identify this discrepancy, noting that "the technical communicator may be a transmitter of messages or a translator of meanings, but he or she is not—or at least not until now—considered to be an author" (p. 80). This reluctance to label technical communication as "authored" communication could come from its connections to the "practical" and "utilitarian" (Miller, 1989). It could also be connected to the fact that the technical communicator is often categorized as an "information manager rather than information generator . . . a glorified typesetter, a font-chooser, or a human grammar check" (Slack et al., 1993, p. 72).

These labels are significant. The way a discourse is categorized is important to our understanding of the power that discourse has in a given community. "At its most mundane," Slack et al. (1993) argue, "certain discourses are granted the privilege of authorship while others are denied that privilege" (p. 80). Using Foucault to examine this claim, these authors come up with something more remarkable: "Rather than authors producing certain discourses [as Foucault would claim in his "author-function"] certain discourses are understood to produce authors" (Slack et al., 1993, p. 80). In the past, members of the field have argued that technical communication is not one of those discourses.

But recasting technical communicators as authors does not solve any problems. If anything, it reinforces the tendency to regard "text" and "author" as finite and delineated entities. To illustrate the paradox, consider the fact that technical communicators (as nonauthors) usually don't get credit for successful communication because the "content" comes from elsewhere. Yet technical communicators (as authors) often do get credit for *unsuccessful* communication (e.g., serving as the locus for punishment in Foucault's author-function.) This paradox is further illustrated in "much of the professional and educational realities of technical communicators" (Slack et al., 1993, p. 85). To the extent that technical communication "education focuses on stylistics, the proper use of forms, and skill at operating the technologies of communication," students are implicitly being taught to be cogs in the wheels of progress rather than agents of change in human organizations (Slack et al., 1993, p. 85). It is this view that contributes to an understanding of writing and communication as individual and isolated processes.

In our (the authors') experiences, some students are quite surprised when they are asked to share "ownership" of their writing with with others. They are often similarly surprised when they find themselves having to navigate the varied and difficult terrain of intellectual property (IP) law when they are just trying to create a website for a client. And while these views are in conflict with many realities of the workplace, they still exist in the "real world." Current legislation certainly makes clear that the "author" (be it corporate or individual) is privileged over the communal or collaborative. Whether this perspective is a poignant iteration of Western values (of ownership and individuality) trumping Eastern values (of cooperation and community) or merely a sad comment on how far we've gotten away from the original purpose of copyright, it remains an interesting challenge as technical communicators enter the global workplace.

Never before have copyright and IP law been so much a part of the everyday writing life of our students. When copyright terms go up eleven times in 40 years, without any real movement forward in our understanding of "authorship" and how technology affects it, we have a difficult problem to address: infinite possibilities to create but limited permission to do so. This is the world that we are training our technical communication students to enter. And we need to teach them what "authority" they have there.

PAST PERSPECTIVES

To understand the legal challenges that face technical communicators in the 21st century, it is necessary to understand the challenges that shaped the profession. Tharon Howard's article (1996) can certainly seem dated in a Web 2.0 world, but only 5 years before Howard asked the question of "ownership," others were asking the question of identity: What should Technical Communication programs do and to whom/for whom should they do it? We would like to think that we have moved well beyond John Mitchell's (1976) view that we "indoctrinate students in the forms appropriate to their employers," for "the students know they must dance with the guy that brung them, and they elect our courses to learn his dance steps" (p. 5); or Susan Wells' (1986) bleak observation that "the ideology of technical writing explicitly assents to its instrumental subordination to capital; the aim of the discipline as a whole is to become a more responsive tool" (p. 18).

Carolyn Miller (1989) does little to assure us that these views are behind us. If anything, she brings the question of practicality (and the attendant questions of practice and pedagogy) to the fore: "Being useful is not necessarily good . . . but little in the discourse of technical writing allows for this conclusion or explores its consequences" (p. 18). To be fair, little in the discourse of the modern, midsized state university would allow for this conclusion either. It could certainly be argued, and argued well, that the university has become a mere reflection of predominant societal values which, in the 21st century (as it was in the 19th and 20th), centers on industry and technology.

The tensions that Dale Sullivan (1990) would address in the winter of the 20th century suggest that this gap between philosophy and practice were still there, even as we turned our eye to the new millennium:

> Those of us who teach technical communication don't often think of ourselves as carrying on the rhetorical tradition . . . indeed, it is rather hard to do, since we teach thought forms and discourse forms demanded by the workplace, and we often find ourselves representing the military-industrial complex rather than the humanistic tradition. (p. 1)

In our current economic climate in which we increasingly have to think about education in terms of efficiency and expediency or production and exchange, it is easier to see the ways in which our field "is indebted to a history that continues to influence our practice and theory, for better or for worse" (Johnson-Eilola & Selber, 2005, p. x). At the same time, "the ubiquity of technical communication is a relatively new phenomenon, one that widens our work in ways before not imaginable" (Johnson-Eilola & Selber, 2005, p. x).

TyAnna Herrington (2001) would usher us into the 21st century by acknowledging the "global proportions" our work has in the digital age, while simultaneously acknowledging the "controlling voices" that seek to narrow that access: "Constructionist ideologies," she writes, "support the belief that communities create knowledge collaboratively, and that the digitization of information and Internet communication provide the metaphorical illustrations of this belief" (p. 114). But the foundational "romantic views of knowledge creation" still challenge the model of virtual communities that "work from collective wealth" (Herrington, 2001, p. 116). Technology has reshaped and advanced our understanding of collaboration, knowledge building, and community, and "the Internet community" has influenced "the way we view intellectual creation" (Herrington, 2001, p. 116). But the law (and the lawyers) are more influenced by the Romantic view of authorship; they "do not allow themselves to reflect on the irony that the Romantic notion of the creative individual is now being exploited to favor the economic interests of media giants at the expense of individual creators, readers, and viewers" (Bolter, 2001, p. xii). We are at a crucial time in our understanding of what is "fair use" and what is "free," what is "property," and what is "community," and how these categories are transformed in the digital age.

AUDIENCE

"Audience" has always been a difficult target to hit. It has not just become so in the world of digital media and social networking. In their article "Audience Addressed/Audience Invoked," Lisa Ede and Andrea Lunsford (1984) identified some boundaries that can still help us today. Essentially, to assume that there is a clearly identifiable audience is as unrealistic as to assume a purely constructed one. The "global village," after all, isn't a real place; even its title functions as a visible reminder of the paradox of modernity, a world where it might be said that "the two great statements of faith are 'What a small place the world is,' and 'I'm a stranger here myself'" (Greene, 1978, p. 89). The global village is "a thoroughly paradoxical place, unified yet multiple, totalized yet deeply unstable, closed and open-ended at the same time" (Ang, 1994, p. 194).

In fact, terms like *globalization*, *multiculturalism*, and *diversity* can semantically undermine the very situations they attempt to depict. As Ang (1994) writes, "The very scope and function of the global village rely inherently on the transmission of homogenizing discourse, unmediated by culture or perspective" (pp. 194–195). Thus, the paradox: Without a shared language, the global village becomes a Tower of Babel—too much of a shared language, and the dominant culture becomes the "global."

This "thingifying" of culture is what Walker Percy (1975) would describe as the "mistaking of an idea, a principle, an abstraction, for the real." The result of such a shift is that "the 'specimen' is seen as less real than the theory of the specimen," and, quoting Kierkegaard, Percy (1975) explains that "once a person is seen as a specimen of a race or a species, at that very moment he ceases to be an individual. Then there are no more individuals, only specimens" (p. 58). Technical communicators need to address people, not just principles. Likewise, teachers of technical and scientific communication need to work with examples and exercises in class that help to complicate, not codify, the idea of "audience." As Ornatowski (1992) observes, perhaps the best way to understand audience is through the lens of ethics—to see "technical writing" as "always rhetorical and involved with potentially conflicting agendas" and to see "technical writers" as "rhetoricians who continually make ethical choices in serving diverse interests and negotiating between conflicting demands" (p. 172). Ornatowski deftly acknowledges that professional writers do not write for "an audience" as much as they write for "many audiences" and that professional writers are not excluded from the audiences they address.

This challenge of understanding and interacting with the world around us is certainly nothing new. Identifying "audience" in a world where the roles have become so blurred, where we are now, more than ever, "among" the audience (Lunsford & Ede, 2009), is made more difficult when the audience is potentially anyone in the world with an Internet connection. In our attempts to teach technical communicators to write for a "broad, international audience," an audience that might be "anyone, anywhere," we risk losing the sense of audience altogether, for "anyone, anywhere" could easily become "no one, nowhere" (Lipson & Day, 2005, p. 5).

Increased access to audiences does not imply increased understanding of how to communicate with those audiences, just as knowing you *can* go somewhere does not mean that you know how to get there. The field of professional communication has begun to recognize this challenge. Most introductory technical communication textbooks have at least a section on "audience" (Anderson, 2007; Lay, Wahlstrom, Rude, Selfe, & Selzer, 2000; Johnson-Eilola & Selber, 2004; Markel, 2010; Peeples, 2003), and others (though fewer) address the complex nature of global audiences (Lipson & Day, 2005; McPhail, 2006). But chapters on audience and globalization can be easily eclipsed by chapters dealing with more immediate, surface-level concerns: style, structure, format, forms—the "old school" way of teaching technical communication.

As Porter (2005) observes, even the STC Code of Ethics proclaims a kind of "client as primary audience" model (253), which can overshadow the larger public/global good emphasis. If the corporation is the client, the "human factor" of audience (be it local or global) is reduced or ignored altogether. The result is a kind of "ethic of expediency" (Katz, 1992) wherein "technological (i.e., economic) expediency is the 'moral' basis of decisions . . . that sometimes harm human welfare or imperil human life" (p. 197). One need only look to the top headlines of 2010—the stream of Toyota recalls and subsequent hearings; the massive explosion at the Upper Big Branch Coal Mine in West Virginia; the catastrophic failure of the BP Deepwater Horizon, belching millions of gallons of oil into the Gulf of Mexico—to see that "technological capitalism" is more important that "technological communication."

From a copyright perspective, Porter's (2005) observation grows even more salient: "As information becomes the new number one product of the US economy, copyright law becomes a crucial scene for playing out the ownership of that product" (p. 252). If the client is both author and audience (and the professional writer merely an instrument "subordinate to capital"), then there is no real concern with communicative action, only commercial interest. The speed with which technology is transforming the public sphere (and the equal speed with which the law is limiting that sphere to protect corporate interest) is something that professional communicators must address. For technology is no longer a tool that stands outside our understanding of the world, it has become the way we understand the world. It has become the world. "Technologies," writes Johndan Johnson-Eilola and Stuart Selber (2005), "are no longer tools to users; they are environments, spaces, worlds, and conversations" (p. x).

AUTHORS

Before we can interact with others in these various spaces, we have to have some understanding of our own position, not only as present-day technical communicators but also as writers who understand how much of our identities were formed from the "heroic self-presentation of Romantic poets" (Woodmansee & Jaszi, 1994, p. 3). Martha Woodmansee and Peter Jaszi (1994) note that "the modern regime of authorship" is not timeless, nor is it universal (p. 3). Rather, it is a relatively recent development, "the result of a quite radical reconceptualization of the creative process that culminated less than 200 years ago" (p. 3). As a result, the Romantic poets of the past tended to view the process of authorship as genuine in terms of it being originary—oringinary, "in the sense that it results not in a variation, an imitation, or an adaptation, and certainly not in a mere re-production"; rather, it was something truly unique and novel (i.e., original) in that it was a "work which, accordingly, may be said to be the property of its creator and to merit the law's protection as such" (p. 3). Thus, these early poetic definitions of authorship form a great deal of the way we as teachers conceive of authorship and the way the law rewards or punishes those who transgress against it.

What is interesting is the degree to which our current conception of authorship implicitly marginalizes communal and collaborative writing practices, practices that "persisted not only in everyday and technical writing," but "among the very writers who contributed most substantially to the Romantic reconceptualization of this activity . . . [Wordsworth's] collaboration with Coleridge is legend, and his extensive reliance on the writing of his sister Dorothy is now also beginning to come to light" (Woodmansee & Jaszi, 1994, p. 3). A more modern iteration might be the ways in which we as teachers value the writing that comes from our classes. Even if we assign collaborative projects, the grading of those projects usually has some form of individualized assessment. In the world of "real-world" collaborative projects, the question still seems to be "Who's on first?"; that is, who is going to get primary credit for this research?

The modern workplace struggles with similar questions. The Council of Science Editors (SCE), for example, published a white paper and a list of references dealing with such subjects as

- The "criteria for authorship"
- The study of "authorship practices" in the sciences (such as "gift authorship" and "name order")
- The question of "(re)defining authorship" in the sciences. (CSE, 1999)

Numerous other studies have focused on the ways in which collaboration takes place in the workplace and examine topics such as are the professional writer and the texts they produce "organizationally situated" (Peeples, 2003). How do we facilitate "productive yet efficient collaboration" in a culture that knows more about "how to win an argument" than "how to exchange knowledge profitably" (Regli, 1999, p. 74)?

As a way to understand and illustrate how these questions manifest themselves in the workplace, the authors interviewed a professional writer who works for a large global company. The purpose of the interview was to explore how some of the theoretical concepts and practical strategies examined in this chapter actually play out in the modern workplace. From this writer, the authors learned that negotiating between individual and corporate culture is the primary challenge in the global marketplace; knowledge of content is secondary (or even tertiary) to knowledge creation, communication, and distribution. As the interviewee explains, "The most important thing is to know how to deal with people. Only then can you do the business of creating the monthly reports that will affect the direction and success of the corporation." And though the "reports are written by several people in several different countries," there is still the question of "who is responsible" for their quality and accuracy; someone performing what Michel Foucault would call the "author-function"—a function that allows a supervisor or vice president to know who to punish if data is incorrect or the report, subpar. And despite all of our social media and new technology, "the main [and preferred] means of communication is the conference call, followed by email." This format allows collaborators from different countries to interact in a way that respects their cultural differences and does not limit them to a homogenized language. As Regli (1999) writes, "We know a great deal and have extensive codification schemes for the rhetoric of persuasion . . . in contrast, we know very little about the rhetoric of interaction" (p. 75). How then do we collaborate and create in the global village without marginalizing or colonizing its inhabitants?

Michel Foucault (2003) addresses the problem of marginalization (and the attendant problems of classification and attribution) in "What is an Author?" His goal in writing this piece was not merely to respond to earlier theories advanced by Roland Barthes (1977) and Jacques Derrida (1998) about the non-necessity of the author. Indeed, his implicit response is that neither critic really gets rid of the author. Instead, they substitute the reader (Barthes) or the text (Derrida) in the author's place, thus reassigning (but not revoking) the systems of power and authority. Moreover, Foucault's goal was not to attempt to pin down what an author "is" (despite his title). Rather, he sought to identify how an author "exists" and "functions" in the culture, ultimately claiming that his aptly named "author-function" is a better way to talk about the complex web of discursive practices that underscore our understandings of "authorship."

In order to give life to his "author-function," Foucault (2003) had to begin by accepting the "death" of the author. He first notes that writing "creat[es] a space into which the writing subject constantly disappears" and goes on to assert that the mere act of writing itself cancels out the writer (p. 378). We write ourselves, as a word or line on a page, precisely because we are going to be absent from the exchange. As such, "the mark of the writer is reduced to nothing more than the singularity of his absence; he must assume the role of the dead man in the game of writing" (p. 378). Foucault's writing here is not simple, but his point, quite so: we read a text in the author's absence. Paul Ricoeur, a contemporary of Foucault, echoes this sentiment in his essay "What is a Text?" when he writes, "Sometimes I like to say that to read a book is to consider its author as already dead and the book as posthumous" (1991, p. 107). And though our definitions of "book" and "text" are certainly changing, Foucault's "author-function" is still a useful way to understand the origins of our understanding of "the author" and to see the opportunities such understanding provides in our global age.

Foucault's author-function is a rich and complex concept, and this complexity allows for a kind of theoretical agility. As a result, the concept functions quite well in describing and analyzing a digitally networked, multinational environment. The author-function, for example, does not refer to a person at all. Rather, it encompasses an entire set of beliefs and assumptions surrounding the "existence, circulation, and functioning" of textual production and consumption (Foucault, 2003, p. 382). Foucault's goal in creating the author-function is to radically "re-examine the privileges of the subject" in order to "deprive the subject (or its substitute) of its role as originator" (p. 390). Because authors should not be imagined as "individualized" and "valorized" heroes isolated from society, they must function as representatives and thus represent a much larger set of socially and culturally constructed ideas. Thus, Foucault's notions of authorship seem well suited to examining different cultural perspectives associated with the creation of text.

This blurring of boundaries between both author and text and between creator and commodity reflects many of the challenges technical communicators face in a globalized, digital age. Admittedly, Foucault is looking at the "author" phenomenon as a kind of post-Renaissance, historical construct used primarily in dealing with literary texts. Yet his conception of the author-function provides criteria by which we can better examine and understand not only our own cherished notions of authorship, but also those ideas held by other cultural groups. Individuals can then use such understanding to publically advocate for change on a local, regional, or public level.

As James Porter (2005) asserted, "Foucault is seldom cited by technical communicators," but he brings "to rhetoric a set of questions that we should adopt in developing an ethics and politics of access and distribution for digital information" (p. 248). For the teacher of technical communication, this aspect means not only addressing the problem of authorship, ethics, and access in our classrooms, but also being prepared to act as agents of change in a post-9/11 world. As Porter explains, "The dynamic of the World Wide Web is being changed, designed and determined by public policy discussions, by legislative and judicial action, and by corporate behavior in which technical communicators are, so far, not much involved" (p. 252). Because these policies directly affect professional practice, there is a need to understand how intellectual property (and copyright) is changing in the digital age and in relation to the increasingly globalized context in which individuals interact.

INTELLECTUAL PROPERTY AND "PROPERTY"

The distinction between physical property and intellectual property might seem to be a rather simple one. Physical property conjures up visions of land, with clearly marked boundaries and with a definite record of who can assert title/ownership. Intellectual property, by contrast, deals with ideas, discoveries, and other products of the mind, all of which are as hard to tie down as air. But regulating the "air" was a part of early American property law. Lawrence Lessig (2004) writes in *Free Culture* that "at the time the Wright brothers invented the airplane, American law held that a property owner presumptively owned not just the surface of his land, but all the land below, down to the center of the earth, and all the space above, to 'an indefinite extent, upwards'" (p. 1). This worked just fine until the technology of airplanes was introduced. Who then had the rights to the air? If it was "ancient doctrine that common law ownership of the land extended to the periphery of the universe" (Lessig, 2004, p. 2), then what space was left for airplanes?

The question was answered, rather succinctly, in 1946 when U.S. Supreme Court Justice Douglas (*U.S. v. Causby*) seemingly erased "hundreds of years of property law" in a single paragraph (Lessig, 2004, p. 2). The crux of the argument relates to how well the overarching idea fits into the context of the modern world. As Douglas argued via analogy, "The air is a public highway, as Congress has declared. Were that not true, every transcontinental flight would subject the operator to countless trespass suits" (as cited in Lessig, 2004, p. 2). Such an argument, Douglas points out, seems contradictory to common sense and creates a situation that could cause problems within such a modern context: "To recognize such private claims to the airspace would clog these highways, seriously interfere with their control and development in the public interest, and transfer into private ownership that to which only the public has a just claim" (as cited in Lessig, 2004, p. 2).

As Lessig (2004) explains, the genius behind this decision lay both in its brevity and in its simplicity: The law should adjust to the technologies of the time, and not vice versa. However, new technologies are increasingly having to be adapted to old laws. Moreover, as the slow wheels of justice start burning down the superhighway, technical communicators have to find acceptable (legal) ways to create in our "permission culture" versus the rapidly evolving digital culture of the 21st century (Lessig, 2004).

Of course, one could claim technical communicators regularly adapt to and address such problems in a variety of contexts. As Killoran (2009) notes, many technical communicators "solve potential IP problems in practical ways that avoid, rather than confront, the disputatious and expensive clarity of legal solutions" (p. 1). But this avoidance can limit the way in which technical communicators individuate themselves in a competitive marketplace. A study by Killoran, for example, found that many participants were afraid to ask clients if samples of their work could be posted once the related project had been completed. "To preclude denial," Killoran says, "several participants mentioned posting or linking only to work that is already in the public domain" or seeking out "work samples that clients would likely find innocuous, choosing a safe topic or an outdated sample of past work, or . . . an unrevealing excerpt such as a table of contents" (p. 5). This example also reinforces the view of the client as uber-author and primary audience; a view that can threaten not only the public good, but the public sphere. What then happens when such ideas and practices are projected into the greater global sphere wherein a range of ideas, cultures, and legal systems come into play? The question is not an easy one to answer. Yet it is an item that requires a response as technical communicators increasingly try to navigate the globalized terrain created by a range of online media.

CURRENT INFORMATION DISTRIBUTION MODELS IN THE GLOBALIZED WORKPLACE

In today's globalized context, there are two opposing models that tend to govern the storing, safeguarding, and distributing of information. One is the traditional model of ownership, which is supported by much of the existing legal framework and emphasizes safeguarding information. The other involves new and emerging models, which are not yet widely supported by the current copyright legislation. (These newer models tend to advance the idea of a more open access to information as a commodity.) Both models influence and are important for the work of technical communicators.

The notion that information has become a commodity in the modern economy is almost a truism. Consequently, access to information might mean profits, while lack of such access might mean failure for many organizations. As Rao et al. (2008) argue, "The protection of intellectual property rights (IPRs) is an essential factor in international economic affairs" (p. 105). Consequently, those individuals who are responsible for the production, sharing, and safe-guarding and organizing of information have become more important in such a globalized economy.

To help individuals better understand such dynamics, Gayton (2008) identifies five main legal issues central to notions of ownership in the globalized workplace. Two of these five items are of particular importance to this discussion. The first is "Intellectual Property Treaty Harmonization," or wider involvement of the new important players in the world market in the negotiation of intellectual property rights treaties and agreements. This process, according to Gayton, has been dominated by the United States and Western Europe until now. Gayton further states that these new regulatory regimes will have to "more accurately reflect [those countries'] 'property ownership cultures' which may be 'in conflict' with the cultures of the US and the European Union" (p. 399).

If Gayton (2008) is correct, then, in a globalized world, both legal professionals and professional technical communicators will have to pay more attention to the legal and cultural specifics of the non-Western environments in which they work. This, in turn, may require future writing professionals to acquire special training in contrastive rhetoric, culture-based norms of collaboration and negotiation, and similar issues.

The second of Gayton's issues to consider is "cooperative censorship," which she defines as the coming together of businesses and governments in limiting access to information (2008, pp. 398–399). According to Gayton, this trend is especially visible in web search technologies in which the developers of such technologies are under pressure to limit users' access to information, thus safeguarding that information. Parallel to this notion, developers of web search technologies might try to "bury" search results in the "bowels of a search and pushing forward paid results regardless of relevance" (p. 399). As most users rarely go beyond the first dozen of hits returned by an online search, this tactic could limit access to the most useful and relevant information for many users (Gayton, 2008).

Rao et al. (2008) present an interesting alternative to this situation. They suggest that, in the computer software industry, other models are being proposed. Specifically, they highlight the notion of "relationship marketing" of computer software—an approach they see as representing a shift from an individual sale-based relationship between the business and the client to building an ongoing relationship with customers (p. 106). According to this perspective, much of the impetus behind relationship marketing in the computer software market came from the burgeoning field of open-source software, which is usually available to users at little or no cost. Among the benefits of such relationship marketing are factors like increased retention of customers and increased brand loyalty (Rao et al., 2008). Rao et al. conclude by suggesting that, in the future, software producers might even make their products available as freeware and only charge for upgrades and support. A somewhat similar business model has been implemented by many Web 2.0 services in which some content and functions are available to users at no change while others can be purchased.

The argument advanced by Rao et al. (2008) seems to imply that traditional models of information and rights management do not always fit into the existing workplace and business realities. It also suggests organizations are looking for new and creative ways to attract and keep customers while also turning a profit. The relationship marketing model and its apparent attitude toward information safeguarding and rights management thus appears to be in conflict with the trend of limiting users access to information highlighted by Gayton (2008).

The two opposing models of managing and safeguarding information as commodity matter to technical communicators for one very important reason. As writers and communicators, technical communicators are in the business of information production, management, and distribution. Therefore, they need to understand how the results of their work are being used and distributed. Moreover, understanding the latest trends and views in information distribution management will enable technical communicators to fit in better with the organization where they work.

TEXT OWNERSHIP AND INTELLECTUAL PROPERTY IN THE WORKPLACE

The concepts of authorship and text ownership in the workplace are often very different from those in academe, where most future technical communicators are trained. This disconnect is further complicated by the fact that relatively few students are exposed to the conventions and ethics of collaboration and borrowing commonly accepted in other cultures, especially non-Western cultures. In the academic settings of many Western cultures, for example, plagiarism and uncredited borrowing are among the worst offenses one can commit. Writers in different workplace settings, however, routinely find themselves either "ghost writing" for others or completing chunks of material written by others into a final product that does not directly attribute credit to these contributors. Similarly, many assessment mechanisms in educational settings value individual work that is easily traceable to an individual writer. The success of a writing project in the workplace, by contrast, is usually measured by its collective outcome and impact on the business's bottom line or organizational goals, not on which member of the writing team contributed the most text to a final product.

Jameson (1993) presents a useful overview of the attitudes toward the "ethics of plagiarism" in business and other professional settings. In most college writing classes, instructors often take the "traditional view" toward plagiarism and documentation and teach students to formally document every source from which they borrow (p. 18). Conversely, writing situations in the workplace might allow (or even demand) a certain effacement of the individual author, thus influencing ways in which authors attribute the material they borrow from others (Jameson, 1993). According to Jameson

> The cosmography of the corporation . . . minimizes the importance of the individual and emphasizes the importance of the collective whole. Each individual employee is expected to contribute to the common good and share in the results. From a legal point of view, all work done—including all written texts—belongs to the corporation. (p. 21)

If Jameson is correct in her assessment of these differences, meticulous attribution of all borrowed or co-written material by corporate authors becomes virtually pointless. As a result, corporate documents usually do not contain such attributions, and many do not even name their authors. Interestingly, citation systems used in academe (e.g., MLA and APA) have made provisions for the existence of the "corporate author." Understanding the difficulty of determining the persons who write professional and business documents, those citation systems simply ask academic writers to list the name of the organization that published a given document.

Reyman (2008) echoes these concerns by stating that despite the "culture of fear" about plagiarizing in academic writing, understanding plagiarism in technical communication "is not that simple" (p. 61). "Technical communicators," she writes, "commonly perform a variety of composing activities [such as using boilerplate materials/templates, relying on existing design formats, ghost-writing, collaborative writing, and cutting and pasting] that could be considered plagiarism in the context of the classroom" (p. 61). Austin (2008), in contrast, notes that "Usually the genres that use borrowed material without attribution can be found within the realm of business and industry, rather than in the academic realm" (p. 609). Austin also argues that determining whether uncredited borrowing is appropriate is situation- and genre-specific. Austin and Jameson (1993) both point to the genre of the business report as one example in which material produced by multiple authors is typically amalgamated into a final document that is distributed to a corporate or an external audience.

TEXT OWNERSHIP, INTELLECTUAL PROPERTY, AND GLOBALIZATION

As Jameson (1993), Reyman (2008), and Austin (2008) all note, plagiarism, borrowing, and attribution are all essentially genre- and situation-specific in nature. If this assessment is correct, then so are the arguments that these practices also depend on the geographical and cultural location in which writing and communication tasks are being performed. In fact, the differences in which various cultures perceive borrowing and text ownership have been well documented by scholars in contrastive composition studies, rhetoric, and professional communication (see, for example, Sechley & Schroeder, 2002; Severino & Salazar, 1997). And as Eva Hemmungs Wirten (2004) points out, much of the current legal intellectual property framework generating the ownership of texts is based on the Western tradition. As a result, other attitudes toward these issues are often ignored. Hemmungs Wirten explains that the relationship between intellectual property and the information society, in many ways, parallels the relationship between wage labor and the industrial manufacturing society of the last century. As a result,

> In the time of spatial interdependence, intellectual property regimes are seen as safety nets designed to protect the capital of our time, a capital which is not only increasingly immaterial rather than industrial, but which more importantly is transborder and oblivious to nation-state regulation. (pp. 10–11)

This assertion raises a key issue for today's technical communicators: In a globalized world, why are we measuring authorship, plagiarism, and information ownership/distribution according to the regulations of a limited number of nation-states (i.e., the U.S. and the EU?) While most scholars agree that we do not yet have solid approaches to this problem, anyone who is involved in cross-cultural and cross-national communication should pay close attention to the implications of the legal uncertainty of today's globalized world. These ideas are not so far removed from those Foucault (2003) raised earlier: Once "discourse" becomes a "thing, a product, or a possession" (instead of a situated "action"), it becomes something able to be regulated, controlled, and policed "by the social order of property which governs our culture" (pp. 382–383). Writing was "a gesture fraught with risks before becoming goods caught in a circuit of ownership" (p. 382)—a gesture perhaps made even more risky as a commodity in the global market.

TYING IT ALL TOGETHER: IMPLICATIONS FOR TEACHING AND RESEARCH

A good deal of work needs to be done in order to bring together academic theory and practice of authorship and text ownership on the one hand and professional practices on the other. We see two directions for this work:

- First, teachers and researchers in professional communication need to reexamine their philosophies and practices when it comes to teaching collaboration, valuing individual and collective contributions to written products and reconsidering definitions of the word *text* itself.
- Second, more effort needs to be put into acquainting current technical communication students with the complex realities of working in the globalized world.

The authors of this chapter see three paths for improving training of students related to issues of authorship and intellectual property in globalized contexts. All three are in some way connected to the teaching and learning of collaboration and project management.

Path 1: Teaching Collaboration Through More and Better Assignments

One of the fundamental issues technical communication instructors at all levels struggle with is teaching students not only to cooperate but to also collaborate on projects. Cooperation typically occurs when students break up a large task into manageable components or tasks and assign these components or tasks to members of a team. As a result of this approach, parts of a final product may be written by group members, but there is little or no subsequent discussion of how these parts even fit together.

Collaboration, on the other hand, is a more complex process that often involves consensus building from disagreement and conflict. This situation is especially true in workplace settings.

One useful approach to teaching collaboration in professional communication courses may be through case studies. Texts like Doheny-Farina's *Rhetoric, Innovation, Technology: Case Studies of Technical Communication in Technology Transfer* (1992), for example, can be useful for many teachers who wish to do more collaborative work in their classes. In the book, Doheny-Farina convincingly shows that the process of workplace collaboration often involves a substantial amount of conflict, with various stakeholders advancing their agendas. He also helps technical communicators learn to see themselves as not only "service workers" to other, more important members of a workplace team but as very important contributors to the whole production process.

Part 2: Valuing Collaborative Projects and Products When Doing Assessment

The traditional method often used to assess students' progress toward a degree in professional communication is the comprehensive examination or a portfolio. And in many cases, such portfolios are expected to contain examples of the work students (as individuals) have done via their coursework in the program. In order to instill a deeper understanding of the importance of collaboration and a new view of text ownership, people in charge of assessing such works should assess students not only on the basis of what parts of the final product or text they actually created individually but also on their contribution to the negotiation and project management aspects of such collaborative projects. As part of this process, students may be asked to submit memos containing self-assessments of their contributions to both product and process.

Part 3: Giving More Attention to Intercultural Aspects of Technical and Professional Communication

Meloncon (2009) notes that only 5% of the graduate programs in technical communication she surveyed required their students to take a course in intercultural or global communication. By contrast, 46% of the surveyed programs required a course in research methods, and 40% required a course in rhetoric. It would be wrong to assume, of course, that students in those programs did not have exposure to issues of intercultural communication in other courses. However, the fact that only a relatively small number of programs required an intercultural communication course of their students is somewhat at odds with the importance of being able to work in today's globalized business environment.

Training in intercultural communication can, of course, take many paths. Some of the most interesting work recently done in this area was by teachers who asked students enrolled in their U.S.-based courses to collaborate with overseas peers on class projects (see, for example, Herrington's 2005 work on the global classroom project).

CONCLUSION

If our research has shown anything, it is that our understanding of author and text, audience and purpose, collaboration and communication (be they local or global) have always been moving targets. As such, the definitions, strategies, and mechanisms we use to address these targets must continually evolve. Similarly, the terrain of intellectual property and copyright, ownership and appropriation is one based on precedent—generally, that of a very narrow range of cultures. As technical communicators, we need to be a part of setting those precedents for the future. We need to make educators, practitioners, researchers, and students aware of the various legal aspects that affect the creation and the sharing of knowledge in global contexts. By understanding past challenges and questions associated with authorship, ownership, and culture, we can better understand the complexity and the opportunities that lie ahead. More importantly, we can use such understanding to educate students and to act as agents of change ourselves.

REFERENCES

Anderson, P. V. (2007). *Technical communication: A reader-centered approach.* Boston, MA: Wadsworth.

Ang, I. (1994). In the realm of uncertainty: The global village and capitalist post-modernity. In D. Crowley & D. Mitchell (Eds.), *Communication theory today* (pp. 193–213). Stanford, CA: Stanford University Press.

Austin, W. (2008). Plagiarism, ghostwriting, boilerplate, and open content. In P. Zemliansky & K. St.Amant (Eds.), *The handbook of research on virtual workplaces and the new nature of business practices* (pp. 604–613). Hershey, PA: Information Science Reference.

Barthes, R. (1977). *The death of the author. Image, music, text* (pp. 142-148). New York, NY: Hill and Wang.

Bolter, J. (2001). Foreword. In T. Herrington (Ed.), *Controlling voices: Intellectual property, humanistic studies, and the Internet* (pp. ix–xii). Carbondale, IL: University of Illinois Press.

Council of Science Editors (CSE). (1999). *White paper: CSE Task Force on Authorship.* Retrieved from http://www.councilscienceeditors.org/services/atf_whitepaper.cfm

Derrida, J. (1998). *The exorbitant question of method of grammatology* (pp. 157-164). Baltimore, MD: Johns Hopkins University Press.

Doheny-Farina, S. (1992). *Rhetoric, innovation, technology: Case studies of technical communication in technology transfer.* Cambridge, MA: MIT Press.

Ede, L., & Lunsford, A. (1984). Audience addressed/audience invoked: The role of audience in composition theory and pedagogy. *College Composition and Communication, 35*(2), 155–171.

Foucault, M. (2003). What is an author? In P. Rabinow & N. Rose (Eds.), *The essential Foucault* (pp. 377–391). New York, NY: W. W. Norton.

Gayton, C. M. (2008). The top five—globalization and the law—predictions for the first half of the twenty-first century. *VINE: The Journal of Information and Knowledge Management Systems, 38*(4), 398–401.

Greene, G. (1978). *The ministry of fear.* New York, NY: Penguin.

Hemmungs Wirten, E. (2004). *No trespassing: Authorship, intellectual property rights, and the boundaries of globalization.* Toronto, Canada: University of Toronto Press.

Herrington, T. (2001). *Controlling voices: Intellectual property, humanistic studies, and the Internet.* Carbondale, IL: University of Illinois Press.

Herrington, T. (2005). Linking Russia and the United Stated in web forums: The global classroom project. In C. Lipson & M. Day (Eds.), *Technical communication and the World Wide Web* (pp. 167–192). Mahwah, NJ: Lawrence Erlbaum.

Howard, T. (1996). Who "owns" electronic texts? In T. Peeples (Ed.), *Professional writing and rhetoric: Readings from the field* (pp. 250–264). New York, NY: Addison Wesley.

Jameson, D. A. (1993). The ethics of plagiarism: How genre affects writers' use of source materials. *The Bulletin of the Association for Business Communication, 56*(2), 18–28.

Johnson-Eilola, J., & Selber, S. (2004). *Central works in technical communication.* Oxford, UK: Oxford University Press.

Johnson-Eilola, J., & Selber, S. (2005). Everything seems to be up in the air at this time: The ubiquity of technical communication. In C. Lipson & M. Day (Eds.), *Technical communication and the World Wide Web* (pp. ix–xii). Mahwah, NJ: Lawrence Erlbaum.

Katz, S. (1992). The ethic of expediency: Classical rhetoric, technology, and the Holocaust. In T. Peeples (Ed.), *Professional writing and rhetoric: Readings from the field* (pp. 183–202). New York, NY: Addison Wesley.

Killoran, J. (2009). *Intellectual property and the web portfolios of technical communication businesses.* Paper presented at the CCCC conference, San Francisco, CA. Retrieved from http://myweb.brooklyn.liu.edu/jkillora/research/2009cccc.pdf

Lay, M., Wahlstrom, B.J., Rude, C., Selfe, C., & Selzer, J. (2000). *Technical communication.* Homewood, IL: Richard D. Irwin.

Lessig, L. (2004). *Free culture: The nature and future of creativity.* New York, NY: Penguin.

Lessig, L. (2010). Getting our values around copyright right. *EDUCAUSE Review, 45*(2), 26–42.

Lipson, C., & Day, M. (2005). Introduction. In C. Lipson & M. Day (Eds.), *Technical communication and the World Wide Web* (pp. 1–23). Mahwah, NJ: Lawrence Erlbaum.

Litman, J. (1994). The exclusive right to read. *Cardozo Arts and Entertainment Law Journal, 13*(29), 29–54.

Lunsford, A., & Ede, L. (2009). Among the audience: On audience in an age of new literacies. In M. Weiser, B. Fehler, & A. Gonzalez (Eds.), *Engaging audience: Writing in an age of new literacies* (pp. 42–70). Urbana, OH: NCTE.

McPhail, T. (2006). *Global communication: Theories, stakeholders, and trends.* Hoboken, NJ: Blackwell.

Meloncon, L. (2009). Master's programs in technical communication: A current overview. *Technical Communication, 56,* 137–148.

Miller, C. (1989). What's practical about technical writing? In B. E. Fearing & W. K. Sparrow (Eds.), *Technical writing: Theory and practice* (pp. 15–24). New York, NY: MLA.

Mitchell, J. (1976). It's a craft course: Indoctrinate, don't educate. *The Technical Writing Teacher, 4,* 2–6.

Ornatowski, C. M. (1992). Between efficiency and politics: Rhetoric and ethics in technical writing. In T. Peeples (Ed.), *Professional writing and rhetoric: Readings from the field* (pp. 172–182). New York, NY: Addison Wesley.

Peeples, T. (Ed.). (2003). *Professional writing and rhetoric: Readings from the field.* New York, NY: Longman.

Percy, W. (1975). *The message in the bottle: How queer man is, how queer language is, and what one has to do with the other.* New York, NY: Farrar, Straus and Giroux.

Porter, J. (2005) The chilling of digital information: Technical communicators as public advocates. In C. Lipson & M. Day (Eds.), *Technical communication and the World Wide Web* (pp. 243–259). Mahwah, NJ: Lawrence Erlbaum.

Powell, P. R. (1999). Facing the audience: Reconsidering "audience" through the Chinese concept of "face." In C. J. Swearingen & D. Pruett (Eds.), *Rhetoric, the polis, and the global village: Selected papers from the 1998 thirtieth anniversary Rhetoric Society of America conference* (pp. 139–146). Mahwah, NJ: Lawrence Erlbaum.

Rao, P. M., et al. (2008). The information and communication technologies and enforcement of intellectual property rights: A relationship perspective. *The Journal of World Intellectual Property, 11*(2), 105–119.

Regli, S. (1999). Whose ideas? The technical writer's expertise in inventio. In T. Peeples (Ed.), *Professional writing and rhetoric: Readings from the field* (pp. 71–79). New York, NY: Addison Wesley.

Reyman, J. (2008). Rethinking plagiarism for technical communication. *Technical Communication, 55*(1), 61–67.

Ricoeur, P. (1991). *From text to action: Essays in Hermeneutics II.* (K. Blamey & J. Thompson, Trans.). Chicago, IL: Northwestern University Press.

Sechley, K. A., & Schroeder, H. (2002). Intellectual property protection of plant biotechnology inventions. *TRENDS in Biotechnology, 20*(11), 456–461.

Severino, R. C., & Carlos Salazar, L. (Eds.). (2007). *Whither the Philippines in the 21st century*. Singapore: Institute for Southeast Asian Studies.

Slack, J., Miller, D., & Doak, J. (1993). The technical communicator as author: Meaning, power, authority. In T. Peeples (Ed.), *Professional writing and rhetoric: Readings from the field* (pp. 80–97). New York, NY: Addison Wesley.

Sullivan, D. (1990). Political-ethical implications of defining technical communication as a practice. *Journal of Advanced Composition, 10*(2), 1–9.

Wells, S. (1986). Jurgen Habermas, communicative competencies, and the teaching of technical discourse. In C. Nelson (Ed.), *Theory in the classroom* (pp. 245–269). Champaign, IL: University of Illinois Press.

Wirten, E. (2004). *No trespassing: Authorship, intellectual property rights, and the boundaries of globalization.* Toronto, Canada: University of Toronto Press.

Woodmansee, M., & Jaszi, P. (1994). *The construction of authorship: Textual appropriation in law and literature*. Durham, NC: Duke University Press.

http://dx.doi.org/10.2190/LIIC9

CHAPTER 9

Software Patent Law in Global Contexts: A Primer for Technical Writing Specialists

Annette Vee

The work and personal lives of technical communicators are often conducted through the medium of software, and so the intellectual property (IP) laws that concern software's composition and international distribution affect the way we write and live. Copyright is the IP regime most often discussed in the field, but patent law has a potentially greater effect on composition in, of, and with software because the monopoly rights granted through patents can be more powerful than the distribution rights afforded by copyright. Patent protection for software has become more readily available through recent court decisions, international treaties, and changes to guidelines for patent examiners in the United States, European Union, and Japan. Software patents, when coupled with our accelerating reliance on both desktop and web-based software, are impacting our lives as writers and teachers.

Several cases illustrate the way that increased patent protection for software has threatened our unfettered access to digital writing and teaching environments. In the high-profile patent case over online content management systems (CMS), *Blackboard v. Desire2Learn* (2009), Blackboard was poised to have monopolistic rights over basic interface features of CMSs before the settlement for the suit created oligarchic rights instead. In another sphere of digital composition—mobile phones—nearly every hardware and software producer is suing every other one (Bilton, 2010). This legal thicket precariously positions what is now both a primary composition space for young people (Lenhart, Arafah, Smith, & Rankin Macgill, 2008) and an increasingly important one for higher education (Johnson, Levine, & Smith, 2009). Most dramatically, the very existence of the

World Wide Web was threatened by a suit from British Telecom, which claimed the company owned all rights to the hyperlink (Jaffe & Lerner, 2004).

Global debates on software patent law are fierce and commensurate with their significant implications for software's composition and distribution. However, the fact that international implementations of software patent law can be complex and arcane may be why it is so seldom addressed in the literature on technical communication or in writing studies more generally. Here, I hope to translate some of the current and global issues in software patents into contexts relevant to writing specialists in order to help us have a better understanding of the ways patent law, especially for software, affects our lives. I begin by outlining some specific reasons patent law pertains to writing specialists, including a more detailed explanation of several patent law cases and an exploration of the blurring boundaries between written text and computer code. I then give an overview of patent law: its relationship to copyright, the ostensible objective of patent protection for intellectual property, what is generally required to obtain a patent, and variations in international patent law regarding software. Finally, I review some of the controversies surrounding patent protection for software, namely, its *de facto* favoring of corporations over independent software developers. Software patents can affect our teaching and composition processes, but they also give us a way of reflecting on our profession as writers and understanding the increasing proximity of our work to the work of computer programming.

HOW SOFTWARE PATENT LAW AFFECTS WRITING SPECIALISTS

The patenting of software directly affects the choices of digital tools we have at our disposal as teachers and writers. Moreover, the disappearing boundaries between writing and coding that patent law implies can lend us insight into our evolving writing practices. Below, I explain in more detail why writing studies must include software patents in our discussions of IP protection.

Direct Effects of Software Patents on Writing and Teaching

Patents directly impact the software environments in which we compose. In a 1991 article arguing against software patents, Garfinkel, Stallman, and Kapor note that XYWrite, a popular word processor at the time, had to eliminate certain features in order to appease another company that had patented them. More recently, OpenOffice has been threatened by Microsoft for infringement of multiple patents (Parloff, 2007). The thicket of patent suits that surround mobile devices (wherein a large part of everyday writing now occurs) is enough to make anyone's head spin (e.g., Microsoft vs. Motorola; Research in Motion vs. Motorola; Apple vs. HTC). Some patents lie dormant (so-called submarine patents), ready to be used against competitors should they be needed to protect the company's business plan. In short, the spaces in which we write for work and play are constrained by software patents and their corresponding lawsuits.

For teachers, this impact is doubled, since software patents affect digital environments for teaching as well as composition. For example, Washington, DC-based Blackboard, the company that makes a popular CMS used at many schools and universities throughout the world, has been involved in at least two major software patent lawsuits, one as plaintiff and one as defendant. In 2006, Blackboard sued the Canadian company Desire2Learn in Texas for infringement of a patent entitled "Internet-based education support system and methods" (USPTO #6,988,138). The patent claimed an educational content management system that would allow multiple users access to multiple courses under different identities (e.g., as instructor, student, teaching assistant).

Despite widespread criticism of the breadth of the patent and educators' attempts to collect information that might lead the patent to be found invalid (e.g., Wikipedia, n.d.), the Eastern Texas district court decided in favor of Blackboard, awarding it $3.1 million in damages (*Blackboard v. Desire2Learn*, 2007). In December 2009, during Desire2Learn's appeal of the verdict, the companies decided to settle their dispute by cross-licensing their patent portfolios (Baker, 2009). The terms of the settlement were undisclosed, as they generally are in cases such as this one, which means that Desire2Learn ceased to challenge the patent and Blackboard's patent still stands.

Continuing the CMS patent wars, TechRadium sued Blackboard in 2008 for infringing a patent that covered a software-implemented system allowing a message to be delivered to various devices when sent from a single, centralized source. This suit also settled privately out of court. As with Blackboard's CMS patent, TechRadium's patent seems quite broad. Moreover, it was originally implemented in the context of emergency notification, yet it potentially extended to *all* software-implemented messaging services that work across different electronic devices. TechRadium also sued the popular online messaging service Twitter for infringement of the same patent in 2009 (Schonfeld, 2009).

These legal battles are fought at great cost to the companies involved. But whether or not they are named in the suit, universities, users of programs, and open-source implementations of software are also caught in the fray. Blackboard's patent suit against Desire2Learn threatened diversity in the CMS market, but it also endangers open-source CMSs such as Sakai and Moodle, as well as any CMS developed locally by universities. To attempt to prevent suits from Blackboard against open-source and university-developed CMS technology, EDUCAUSE banded together with universities. In response to this pressure, Blackboard pledged to not use their patent against noncommercial CMSs, thereby assuaging a rapidly escalating public relations nightmare with their main clients (Carnevale, 2007). While this pledge offers some comfort to those who wish to use a non-Blackboard CMS, the pledge is nonbinding—it does not negate the patent itself nor Blackboard's right to sue anyone implementing a system that might infringe it. The mere existence of a broad-sweeping patent so central to online learning is likely to discourage innovation in this educational sector.

If a multiple-login CMS system such as Blackboard's can be patented, what other methods and programs that we use to teach might be patentable? How might this affect the ways that we use technology to augment education? The vast number of software and methods patents, the high variability in how they are treated across the globe, and the increasing popularity of distance education makes the answers to these questions impossible to determine but important to explore.

Disappearing Boundaries between Writing and Coding

Writing online or with software documentation is common in professional writing; we write with and for websites, APIs (application programming interfaces), and online spaces such as Twitter and Facebook. Many technical writers are also scripters or programmers for these interfaces. As the line between composing software and composing more traditional texts is growing fainter, the legal fate of software may hold keys to the fate of writing as well.

In 2000, the use of the hyperlink—perhaps the clearest example of hybrid text/code common in technical writing—was threatened when British Telecom sued Prodigy for infringement of the patent they claimed to hold on it (Jaffe & Lerner, 2004). Since the hyperlink is a basic building block for the World Wide Web, British Telecom's patent threatened all uses of writing on the web. The suit was finally thrown out, but as Jaffe and Lerner (2004) argue, "What is so scary about this case is that British Telecom's assessment of the current state of the patent landscape gave it sufficient hope of prevailing so as to make this case worth bringing" (p. 58).

Even when professional writers are not coding alongside writing, we often work closely with programmers and in similar compositional contexts as they do. Oracle's 2010 purchase of Sun, the originator of the open-source language Java, alarmed many employed and freelance programmers because of Oracle's far more aggressive reputation for prosecuting their IP (e.g., Kuhn, 2010). Sun had pledged not to enforce their patents on Java, allowing programmers to improve the language through open-source methods; however, Oracle is not bound by Sun's promises to the open-source Java community. Software (much of it written in Java) defines the delivery of our work and shapes its composition; the fate of our profession may be mirrored in the fate of programmers alongside whom we work.

The blurred line between software and technical writing extends to teaching as well. Alfred Essa (2007) warns that software patents will become increasingly relevant as well as restricting as higher education shifts teaching toward online spaces, for

> As the world becomes increasingly digital, more educational technology tools become commonplace, as more faculty utilize cyberspace for some part of their teaching, and as the boundary between teaching in the traditional classroom and teaching online begins to blur, our intuitions about property, collaboration, and academic freedom will be put to the test. (p. 72)

From his perspective as Associate Vice Chancellor and Deputy CIO for the Minnesota State Colleges and Universities, Essa sees higher education under attack from politics, for-profit institutions, and decreased public funding. "Information technology is one of the few strategic levers we possess for controlling costs and improving quality," he argues; yet "the specter of software patents has now placed that strategic lever at risk." The fact that software—processes embodied in texts—can be patented suggests that even teaching methods could be subject to patents.

OVERVIEW OF PATENT LAW

Now that I have outlined specific reasons for professional writers to take note of trends in patent protection for software, I will explain a bit of the history and nature of patent law to protect intellectual property, especially as it pertains to software. I focus primarily on the United States and the European Union; along with Japan, these jurisdictions account for the vast majority of software patents issued worldwide (Park, 2005). This explication is not designed to be comprehensive, but it provides sufficient background to understand various issues in patent law that might affect technical communication and writing studies more generally.

Patent Law vs. Copyright Law

Since copyright law is likely to be more familiar to professional writers, we can look at patent law's relationship to copyright to understand how it functions as intellectual property protection. Copyright protects original expressions and is an especially important form of protection for writers, musicians, and artists. Patent rights protect "inventions" that are functional and are more central to the work of engineers, architects, and chemists.

Distinctions between function (generally protected by patents) and expression (generally protected by copyrights) have proven very difficult for courts to determine. For example, in *Brandir International, Inc. v. Cascade Pacific Lumber Co.* (1987), at issue was whether an elegantly designed yet functional bicycle rack could be afforded copyright protection. The Court decided that because its function could not be fully separated from its aesthetic form, the "Ribbon Rack" was not afforded copyright protection. For writing, however, copyright law has been able to draw a distinction between expression and function. In the landmark case *Feist Publications v. Rural Telephone Service* (1991), the U.S. Supreme

Court asserted that the mere collection of facts was below the bar of "original expression" in U.S. copyright law. In contrast to the phone book under discussion in *Feist*, instruction manuals, photo captions, cooking directions, and business information on websites are all subject to copyright because they involve a kind of creative expression.

Both copyrights and patent rights are routinely assigned to people or institutions that are not the creators; pharmaceutical companies often own the patents on the work of their chemists, and publishers often own the copyright on the writing they publish. While patents generally bar anyone from manufacturing or re-creating an invention without the patent holder's permission, copyrights prevent others from copying the copyright holder's work. To infringe a copyright, then, a person must have had access to the original work to copy it. Since proof of patent infringement does not require that the alleged infringer had access to the original invention, however, a person can violate a patent of which she is unaware.

Like the Ribbon Rack, software is a difficult nexus of function and expression because it is both (aesthetic) text and (functional) virtual machine. Software is composed of source code, a specialized kind of text written by programmers to direct the computer's processes. Source code is writing that turns a computer into a virtual, specialized machine, and so it can be thought of as a text, a machine, or both. When framed as a text, source code is subject to copyright protection. When framed as a machine, it is subject to patent protection. Whether it is in the form of object code (the code that the machine can read) or compiled code (the human-readable language in which it is written) or whether it is in an interpreted programming language (i.e., one that does not require compiling), source code is assumed to be subject to copyright. This presumption of copyright protection for source code is global, as various international treaties such as the Berne Convention harmonize copyright protection. The presumption of patent protection for source code, or the software that is constructed from it, is more controversial and is not uniformly implemented through international IP law.

What is Patentable?

A patent is a government-sanctioned, limited monopoly on the manufacture or practice of a particular invention. Specific requirements that an invention must meet to be patentable are up to individual nations to determine, subject to some international treaties and organizations, but common requirements include specific subject matter, novelty, usefulness, and disclosure. To obtain a patent, an inventor must apply to a national authority that grants patents. If it meets a nation's particular requirements, she can obtain a patent and exercise her limited monopoly for the time specified by that nation.

The monopoly right of a patent is limited by time: at least 20 years for all signatories to TRIPS (Agreement on Trade-Related Aspects of Intellectual

Property Rights), an international treaty governing patents and that includes the United States, Europe, Japan, and all other World Trade Organization (WTO) countries. Theoretically, patents encourage innovation by allowing the inventor to recoup her investment in researching and developing the invention. Yet the disclosure of the invention (i.e., the full description of it) allows others to learn from its development and eventually to practice or make the invention once the patent expires. This exchange of information is central to the economic theories of patent law.

Requirements for Patent Claims

Subject Matter

Subject matter available for patenting differs across nations, especially for software. "Inventions patentable" is defined by 35 U.S.C. §101 as "Whoever invents or discovers any new and useful process, machine, manufacture, or composition of matter, or any new and useful improvement thereof, may obtain a patent therefore, subject to the conditions and requirements" specified in subsequent Sections defining the conditions of novelty, nonobviousness, and disclosure. The European Patent Organization (EPO), which provides guidelines for EU patents, outlines what is and is not subject to patent protection: "European patents shall be granted for any inventions, in all fields of technology, provided that they are new, involve an inventive step and are susceptible of industrial application." Explicitly excluded as inventions are "rules and methods for performing mental acts, playing games or doing business, and programs for computers" (European Patent Convention, 2000). Japanese patent law states that an invention is patentable if it is "a highly advanced creation of technical ideas utilizing natural laws" (as cited in Park, 2005). Game rules, economic principles, and human mental activities are not patentable in Japan because they rely on human situations rather than natural laws (Park, 2005).

As suggested in the terms *technology* (Europe), *invents or discovers* (U.S.), and *creation* (Japan), patentable subject matter generally excludes anything already found in nature or anything not invented by humans. What is *natural,* however, can be difficult to define. Medicines derived from plants, genes (when created or processed by humans), and applied mathematical algorithms have been eligible for patenting worldwide, although they are all arguably found in nature.

Software is explicitly excluded from patentable subject matter in Article 52, 2(c) of the European Patent Commission; however, the EPO does issue patents on "software-related inventions" (Nard, Barnes, & Madison, 2005, p. 759). Jinseok Park (2005) provided cases that illustrate the distinctions the EPO makes between "software as such" and "software-related inventions." For example, an EU court asserted, "If the program controls the operation of a conventional general-purpose machine so as technically to alter its functioning, the unit consisting of program and computer combined may be a patentable invention" (*Koch & Sterzel* T 0026/86-3.4.1, as cited in Park, p. 338). Additionally, in *Vicom*, the Boards declared that a mathematical method, when directed at a technical process, was patentable (EPO Decision T 0208/84-3.5.1, as cited in Park, p. 338). The Japanese Patent Office (JPO) treats software similarly to the EPO (Park, 2005).

Compared to the JPO and EPO, the USPTO has been more liberal in issuing patents on software. This is especially true since *State Street Bank and Trust v. Signature Financial Group* (1998), when the U.S. Court of Appeals for the Federal Circuit (CAFC) lifted the case law exemption for the patentability of business methods. Prior to that, patents on software were issued by the USPTO, but they had to be embodied in a machine and their status was more ambiguous. *State Street* indicated unambiguously that computer programs *per se* were patentable as long as they were useful. As a 2000 EU study on the patentability of computer programs states,

> The real difference between the USA and Europe is that in Europe the invention has to be of a technical character whilst in the USA the mere fact that the invention uses a computer/software makes it of the technological arts, if also useful, concrete and tangible results are provided. (Hart, Holmes, & Reid, 2000, p. 2)

However, following the *State Street* decision in the United States, the EPO began to issue many more patents on business practices as well as data processing and equipment (Blind, Edler, & Friedewald, 2005). These disparities in patentable subject matter mean that computer-related inventions are protected differently across nations despite the fact that many are globally distributed through the Internet.

Novelty and Nonobviousness

To meet the novelty and nonobvious requirements for patenting, inventions generally must demonstrate some kind of innovation; one cannot patent a wheel, for instance, since it is technology widely known and used. However, a particular type of wheel, constructed in a way to serve a unique purpose, may be eligible for a patent if it involves "an inventive step" (see TRIPS, outlined below). If the step is not "inventive," then the invention might be considered obvious in light of available technology and therefore ineligible for patenting.

The novelty requirement for U.S. patents is specified in 35 U.S.C. §102(a): if, previous to the patent-applicant's invention, "the invention was known or used by others in this country, or patented or described in a printed publication in this or a foreign country" then it is not novel. In other words, the United States does not consider foreign inventions to be prior art if they are not patented or described in print. In contrast, public foreign knowledge *is* treated as prior art by the EPC (Article 54(2)) and Japan Patent Law (Section 29(1)) (Nard et al.,

2006, p. 700). This treatment of foreign prior art means that U.S. law allows for U.S. corporations to patent such things as foreign indigenous plant remedies and profit from them in the United States, sometimes called "biopiracy."[1]

Utility

The utility requirement of patents is expressed by 35 U.S.C. §101 ("useful") and Article 54 EPC ("industrial applicability"). Seldom does the lack of an invention's utility bar its patenting because patent offices and courts often assume that utility can be more clearly determined by market forces than by law. When utility does emerge as a challenge to a patent, it tends to be in the field of biotechnology, when a company attempts to patent something that appears to be a preliminary stage of a research trajectory (Nard et al., 2006).

The court deciding the case *Brenner v. Manson* (1966) articulated one value of the utility requirement: the *quid pro quo* patents provide between inventors and society through the sharing of useful technology (Nard et al., 2006). As the *Brenner* court argued, technology should not be patented prior to its moment of utility because a patent on an unfinished process "may confer power to block off whole areas of scientific development, without compensating benefit to the public" (as cited in Nard et al., 2006, p. 737). The hamstringing of innovation that the *Brenner* court warns about may be especially pertinent to software composition because of its rapid iteration and development cycle.

Disclosure

The *quid pro quo* of patent law is most clearly enacted in the disclosure requirement. In exchange for a patent, an inventor must explain how to make or practice the invention in the patent's "specification." Companies naturally prefer to be as circumspect as possible when describing their inventions in order to keep their innovations from their competitors. Yet without the full description of the invention, the patent holder would receive a monopoly right without contributing to the public good. Consequently, nations generally specify the level of detail required to obtain a patent. For example, the United States requires enough detail for a "person having ordinary skill in the art" of the field (the PHOSITA) to reproduce the invention.

The disclosure requirement has been oddly applied in the case of software in the United States, and some critics suggest that too many junk patents on

[1] Nard et al. (2006) described a 1990s case in which a multinational company, W. R. Grace, patented products made from the Indian neem tree. These derivations of the neem tree were widely used by people in India, but they had not been patented. When Indian NGOs challenged Grace's patents, the EU patent was invalidated for lack of novelty, but the U.S. patent stood, allowing W. R. Grace to continue their government-granted monopoly on Indian folk uses of the neem tree within the United States (p. 700).

computer programs have been issued in part because the specification requirements are low. One of these critics, Ben Klemens (2006), describes several software patents in which the requisite "written description" includes only vague flow charts rather than the code that would pin down the methods more concretely. In *Northern Telecom vs. Datapoint Corp* (1990), Northern Telecom showed evidence that it was not necessary to reduce their software invention to code in their patent specification because their written description differed only slightly from the actual code that enacted it. The Federal Appeals Court agreed, suggesting that the translation of software design into code is merely a technicality (Burk & Lemley, 2005).

The U.S. Supreme Court says that software is "ready for patenting" once a commercial order is placed and the rough outlines of the software are drawn, even if the code is never written or does not work (Burk & Lemley, 2005). However, as technical communicators can attest, the difference between an abstract or general description of a proposed text and the actual text can be significant. Through this lens, U.S. patents on software come dangerously close to patents on *ideas*, which are not allowed in any nation.

Patent Law Jurisdiction

Jurisdiction for patent law is at the national level. For example, the United States stipulates its own requirements for issuing patents, the USPTO issues patents, and the U.S. court system handles disputes about U.S. patents. Since a patent is only enforceable within national boundaries, a person could not be barred from making a U.S.-patented invention in Japan and selling it there unless a Japanese patent is also obtained for the invention. In the EU, member states issue their own patents, but the EPO also offers patents and provides guidelines to member states. If an inventor obtains an EU patent, she must have that patent validated by each member state in which she would like protection.

Organizations such as the WTO (World Trade Organization) and WIPO (World Intellectual Property Organization) and treaties such as TRIPS (Agreement on Trade-Related Aspects of Intellectual Property Rights) stipulate how patents must be honored beyond national boundaries. Because TRIPS is attached to the WTO, enforcement of the agreement is possible through trade sanctions (Mueller, 2009). TRIPS codifies requirements for patenting: "Patents shall be available for any inventions, whether products or processes, in all fields of technology, provided that they are new, involve an inventive step and are capable of industrial application" (as cited in Mueller, 2009, p. 544). Countries that wish to become WTO members and TRIPS signatories must have patent laws that reflect this requirement, although they have leeway in how they interpret that clause. TRIPS explicitly mentions that copyright law pertains to computer programs, and it calls the computer programs "literary works," to be treated as specified by the Berne Convention, which pertains to copyrights. Therefore, computer code is almost universally covered by copyright law, but its coverage under patent law varies across nations.

Numerous minor agreements impact patent law as well. One example is the Paris Convention, which harmonizes the file date for patents; if an inventor files for a patent in Japan, then files in the United States, he gets the earlier Japanese filing date for review in terms of novelty and nonobviousness (Mueller, 2009). Language barriers, complex enforcement of international treaties, and differences in patentable subject matter and requirements across nations can make the process of enforcing patents internationally much more complex and expensive than enforcing them nationally, which is already very difficult without the help of patent attorneys.

Shifts in International Patent Law

Intellectual property laws have generally been stronger and more established in Europe and the United States than in other countries. Beginning in the 19th century, as trade and transportation accelerated communication across the globe, IP law radiated out from the centers of industry and began taking on an international scope. For example, the Berne Convention harmonized copyright for literary and artistic works among many nations in 1886 (although the United States did not sign on until 1988). In the last few decades, IP law harmonization through international treaties has tended to increase protection throughout the globe. TRIPS was negotiated through the WTO in 1994 and is largely the result of pressure from more industrialized, IP-producing nations like the United States, Japan, and members of the EU.

Further evidence of global trade and efforts to harmonize patent law to facilitate it can be found in the Five IP Offices working group. The Five IP Offices is a group organized to streamline processes of patent filing in the largest patent offices in the world: Japan, Korea, China, Europe, and the United States, which together account for 90% of patent applications filed worldwide (Five IP Offices, n.d.). The Five IP Offices, whose heads met in 2008, claim they want to streamline their applications because "as the world sees economic barriers between nations fade away, innovators want their intellectual creations protected concurrently in multiple major markets" (Five IP Offices, n.d.). The USPTO and the EPO are also currently working on a joint classification system that will make it easier for companies to file for patents in both the EU and the United States (USPTO and EPO work, 2010). This international focus on trade is corroborated by patent law scholar Janice Mueller (2009), who noted that international patent law legislation in the 1980s shifted from the WIPO, which concentrated on IP, to the WTO, which covers international trade more generally and seeks to lower trade barriers between nations.

Although the focus of this chapter is not on international trade relations, it should be noted that lowered trade barriers often imply imbalances in power between industrialized and developing nations. To illustrate, language about what is patentable under TRIPS was aimed at less-developed countries who were refusing to grant patents on agricultural and pharmaceutical inventions because monopolies on those inventions increased their price, sometimes prohibitively (Mueller, 2009). Under TRIPS, some compulsory licensing is allowed for desperately needed pharmaceuticals and agricultural technologies, but only with certain conditions met: a national emergency or extreme situation must exist; reasonable terms for a consensual license are not available from the patent holder; and the patented invention must be for use only in the domestic market (Mueller, 2009). This ethical issue of technology transfer between industrialized and developing nations is also addressed indirectly in Article 52 EPC, which provides an exception to patenting to protect "*ordre public*" or "morality," suggesting that such technology as highly productive varieties of grain or AIDS drugs should be made widely available.[2]

In the United States, the general trend toward increased copyright protection has moved in parallel with increased patent protection. Although legislation has driven increased copyright protection (e.g., the Digital Millennium Copyright Act), most of the changes related to software patents in the United States have occurred through shifts in policies at the USPTO and major decisions in the courts. For example, the Court of Appeals for the Federal Circuit (CAFC) was established in the early 1980s to streamline the patent appeal process, and it has interpreted patent law to make it easier to get patents, easier to enforce patents, easier to get large financial awards from such enforcement, and harder for those accused of infringing patents to challenge a patent's validity (Jaffe & Lerner, 2004). Application fees for patents finance the USPTO and also provide a surplus of discretionary funds that Congress uses to fill budget shortfalls elsewhere. This budgetary reliance on patent fees translates to governmental and USPTO support of patent issuance. As Jaffe and Lerner (2004) explain, the concern about American competitiveness in the global marketplace in the 1980s led Congress to push policies that turned the USPTO and patent system into more of a business than a government organization.

Software Patent Law

When software first became important to businesses in the 1960s, the prevailing opinion among intellectual property scholars in the United States was that

[2] However, Nard et al. (2006) noted that despite some member states' arguments along morality lines, the EU has allowed patents on human DNA (e.g., the hormone Relaxin, used during childbirth). Moreover, the United States, the origin of many of these technologies and patents, does not have a stipulation for "*ordre public*" or "morality" as part of their patent code. The work of Vandana Shiva (2001) further illustrates how patents held by large international companies such as Monsanto can cripple agriculture in developing countries.

software, algorithms, and business methods were too abstract or too much like pure math to be patentable (Merges, 1999). President Johnson assembled a special commission of academics, scientists, and industry representatives in order to review the USPTO's stance on patenting software; the commission ultimately recommended that software should not be patentable (Samuelson, 1990). Court cases in the 1970s affirmed the idea that computer programs were too abstract to be patentable (*Gottschalk v. Benson*, 1972; *Parker v. Flook*, 1978). The Freeman-Walter-Abele test, the case law standard for determining patentability of computer-related inventions at that time, indicated that an invention claiming a mathematical algorithm had to have that algorithm applied to a physical process in order to be patentable.

Patent cases in the 1980s began to chip away at the special exemptions for algorithms and business methods. *Diamond v. Diehr* (1981) indicated that although an algorithm *per se* was not subject to patent, a specific application of an algorithm in an industrial process was potentially patentable. After this case, patent claims emphasized the "industrial process;" they began to use the phrase, "a general-purpose computer on which is programmed a method to calculate" rather than "a method to calculate" (Klemens, 2006, p. 45). *In re Alappat* (1994) further undermined the software patent exemption. In that case, the CAFC wrote, "We have held that such programming creates a new machine, because a general purpose computer in effect becomes a special purpose computer once it is programmed to perform particular functions pursuant to instructions from the program software" (as cited in Klemens, 2006, p. 58). In other words, a new software program *legally created* a new physical machine.

The case that removed all doubt about the patentability of software in the United States was *State St. Bank and Trust Co. v. Signature Financial Group, Inc* (1998). In *State Street*, the court unambiguously declared that the "so-called 'business method exemption'" to patentability was invalid. In considering patentability, the emphasis should not be on subject matter (specified in 35 U.S.C. §101) but on the result: as long as a process has a "useful, concrete and tangible result," it is patentable subject matter (as cited in Nard et al., 2005, p. 757). Because software was considered a kind of business method, a flood of applications for software patents inundated the USPTO after *State Street*.

Software is still considered patentable subject matter in the United States, although more recently, the U.S. Supreme Court rejected the "useful, concrete and tangible result" test in favor of the "machine-or-transformation" test, which states that a process claim is eligible for patenting "if: (1) it is tied to a particular machine or apparatus, or (2) it transforms a particular article into a different state or thing" (*Bilski v. Kappos*, 2010, p. 1). Although critics of software patents had pinned hopes on the Supreme Court's review of *Bilski*, the Court's decision in *Bilski* was narrow, and they "decline[d] to adopt a broad exclusion over software or any other such category of subject matter beyond the exclusion of claims drawn to fundamental principles set forth by the Supreme Court" (p. 21 n. 3).

Other countries can resist the U.S. model of increased software patent protection because TRIPS does not force signatories to consider software patentable subject matter. For instance, New Zealand was considering a ban on software patents (Heras, 2010). Moreover, the EU policy is currently applied heterogeneously across member states, since each has the power to grant its own patents and implement its own interpretation of EU patents. Support for software patents is stronger in the United Kingdom than in other member states, for instance.

A commission was formed in 2002 to review the EU policy on software patents and consider whether the U.S. approach would be economically beneficial. This commission specifically considered the impact of patents on open-source software, which contributes substantially to the technological economy in the EU (Blind et al., 2005). The commission was also charged with the goal of homogenizing support for software patents across EU member-states in order to lower some of the financial and bureaucratic barriers to technological business in the EU. During the review process, software patents were hotly debated among technology experts, industry veterans, and intellectual property scholars. Across the globe, industries, lawyers, and programmers watched with interest as Europeans considered whether to adopt a U.S.-style system or to continue to exempt software from patentable subject matter. The final vote in the EU Parliament rejected the patentability of software programs *per se* (Patentability of computer-implemented inventions, 2005).

In summary, patents are generally awarded for human inventions that are useful, nonobvious, and genuinely new, although the specific treatment of software as patentable subject matter varies worldwide. Like copyright law, patent law is subject to interpretation by individual nations, but it is affected by numerous international treaties. Intellectual property law regarding software is particularly susceptible to international laws because its location is often hard to pin down to a particular national jurisdiction. And yet software patents affect all users of software worldwide. In the final part of this chapter, I consider what the controversies of software patent law have to do with writing studies specifically.

CONTROVERSIES SURROUNDING PATENT PROTECTION FOR SOFTWARE

Given the intimate connections between writing and software, we should be concerned with the balance of legal protection awarded to software. Legal protection for intellectual property is designed to stimulate new production of ideas and technology for the benefit of society. Yet when protection is too strong, it can work instead to keep ideas and technology out of the hands of the public. As Lawrence Lessig (2002) has argued, stronger intellectual property protection does not necessarily lead to greater opportunities for innovation. If patent protection encourages innovation and diversity in software, then we should encourage it; conversely, if patent protection constrains our choices in teaching and composing in software, then we should be wary of that protection. Unfortunately, the benefits and drawbacks of patent protection of software are not straightforward or clear. With a goal of helping us to understand this ideal balance to encourage innovation, this section looks at the arguments and evidence surrounding patent protection for software.

Opposition to Patents by Software Writers

Patent protection was designed within a paradigm of industrial production—highly centralized development with significant investment in research and development. Because of software's unique development profile, many developers argue against any kind of patent protection for it. They argue that patents are not as good a fit for software as they are for industrial technology; they can tie up ideas or bracket off potential innovation more so than in other fields. In contrast to biotechnology, engineering, and industrial production, the field of software production often has low start-up costs and rapid development and iteration. An independent software developer can produce patentable technology using no resources other than a computer. This also means that the independent software developer working on her home computer can *violate* patents, most of which are held by companies with significant legal resources such as IBM and Microsoft.

Blind et al. (2005) showed that although some companies had benefited from a software patent, this benefit was not seen across firms on average. Moreover, firms that did not use software patents expressed uneasiness that their actions were restricted by software patents. The authors indicate that the differences in software development—shorter development cycles, frequent new products, and need for interoperability—make patents less beneficial for that field. Other dangers of patent protection peculiar to software include the collaborative nature of software composition, the difficulty of searching for prior art (Blind et al., 2005), and the sheer volume of code written for software (Garfinkel et al., 1991).

Perhaps the most vocal and long-standing critic of software patents has been Richard Stallman, author of the popular General Public License (GPL), which is used prominently in open-source software development. In an article he coauthored with Simson Garfinkel and Mitchell Kapor (1991), Stallman argued that the USPTO's standards for awarding software patents was too low and that (even before *State Street*) the flood of patents on "ideas so elementary that they could have been answers to problems in a first-year programming course" (p. 51) did real harm to the innovation in software. The authors note that large companies such as IBM, Lotus, and Microsoft became successful without software patents. Lawrence Lessig (2002) points out the odd fact that patent protection for software in the United States came not from arguments by the software industry but from the courts; major software companies Adobe and Oracle both came out against software patents in 1994 because they did not see them as useful incentives to software innovation.

Computer game pioneer and id Software founder John Carmack, who often releases game code under Stallman's GPL license, has repeatedly refused to patent his own programming techniques because he believes that patents on software slow innovation and creativity (Carmack, 1997). Other entities who release free software under Stallman's GPL license—many of them universities or individuals—cannot afford the risk of patent infringement. Software patents also do not reduce trade secrecy, which is one aim of the patent system, because the disclosure requirement for software is relatively lax (Garfinkel et al., 1991). Many developers argue that the intellectual property laws of copyright and trade secret protection fits the development profile much better than patent protection.

Opposition to software patents is particularly strong among independent software developers; in a survey of European software developers, independent developers were much less likely than larger software companies to support stronger and more globally binding patent law (Blind et al., 2005). They believe software should be excluded from patenting, and nearly all surveyed strongly disagree that Europe should move to a model such as the United States has, with patentability for software as such (Blind et al., 2005, p. 87). Tech-centered Internet forums such as Slashdot also demonstrate the popular sentiment against software patents among individual developers; see, for instance, the community's uproar over the patent-supporting brief that IBM submitted to the U.S. Supreme Court during the *Bilski* trial (IBM's Supreme Court Brief, 2009).

The Problems of Patent Protection for Independent Software Development

The composition and distribution practices of independent software developers are part of why they are more likely to oppose software patents. An extensive survey conducted in 2008 by prominent U.S. patent law scholars revealed that especially for software start-ups, "patents provide relatively weak incentives for core activities in the innovation process, such as invention and commercialization" (Graham, Sichelman, Samuelson, & Merges, 2009, p. 1261). As a result, small software companies are much less likely to patent their technology. On the other hand, companies that consider the core part of their business to be product innovation (those that mainly make hardware or physical technology) were twice as likely to indicate that patenting was "important in capturing competitive advantage" (p. 1293). Software companies, even those who make software "products," or packaged software for sale, are less likely to see patents as central to their competitive advantage; instead, "first-mover advantage" is much more important (Graham et al., 2009, p. 1293).

Blind et al. (2005) found similar tendencies against patenting for smaller software firms in Europe.[3] They showed that the tendency to patent increases as the size of the corporation increases. An increased research budget did not correlate with higher patenting for software, implying that the availability of patents for software does not promote research investment. Like Graham et al. (2009), Blind et al. (2005) also found that the tendency to patent was higher in hardware-centered corporations, showing that the patenting profile of hardware-plus-software firms followed the higher-patent model of hardware rather than the lower-patent model of pure software firms.

Noting that smaller firms lack expertise in patent law and have more limited budgets, Blind et al. (2005) indicate that these firms have "neither the awareness nor the resources to make effective use of patenting" (p. 19). Indeed, Graham et al. (2009) show that small software firms are more sensitive to the costs of patenting and prosecution. They cite that "the average out-of-pocket cost for a respondent firm to acquire its most recent patent was over $38,000," and smaller firms not only have fewer monetary resources, they also need to get out-of-house counsel, which is more expensive and harder for them to monitor (Graham et al., 2009, p. 1311). Lessig (2002) points out that the marginal costs for smaller firms to patent software is much greater than it is for firms with much larger budgets.

Compounding the problem that independent developers have lower *de facto* access to patents because of their lack of resources, independent developers tend to be disproportionally affected by software patents. Law regarding patent infringement does not discriminate between large corporations and individuals and so the lack of resources among independent developers can lead not only to their lack of patents but also to their greater vulnerability to patent infringement suits. As a U.S. court stated in *Lough v. Brunswick Corp.* (1996), "The law does not waive statutory requirements for inventors of lesser sophistication" (as cited in Nard et al., 2006, p. 685). In other words, everyone must play the game the patent lawyers play. In fact, independent developers are significantly more likely to have a project impeded, go up in price, or prolonged because of software patents, and the threat of these effects mean that they are less likely to start a project in the first place (Blind et al., 2005). They may also be more likely to hide their code or keep their business small to stay under the radar of patent-infringement suits (Vee, 2010).

Open-source software development is even more vulnerable to harm from software patents. Because other incentives exist for open-source software development, patents do not do much to encourage innovation in that sector, yet the development process is still subject to patent infringement (Blind et al., 2005; Vee, 2010). Even if an open-source software group wanted to patent their work,

[3] Although patents on software per se are not supported in Europe, this study focused on *software-related* patents.

their marginal cost of obtaining a patent would be magnified because budgets in open source are generally nonexistent. Plus, the organization of open-source development is so decentralized that a coordinated effort toward patenting is nearly impossible (Lessig, 2002). Many commercial developers and other industries rely on open-source software, so the negative effect of software patents on open-source development bleeds out into the commercial sector as well (Blind et al., 2005). A dearth of economic studies on the impact of open-source software means that we cannot know the full extent of the damage.

In a global context, the negative effect of software patents on small-scale, independent, and open-source development could be accentuated. The patchwork of regulations makes it difficult to navigate patents internationally; firms with less access to monetary and legal resources will find it much more difficult to patent their technology in international markets and yet will be more vulnerable to the threat of patent infringement lawsuits. International patent law allows for mandatory licensing on such technologies as pharmaceuticals for health, but this can reify the consumer position of developing countries. Although he does not address patent law directly, an extension of Arjun Sengupta's (2004) argument for a "human right to development," based on a 1986 United Nations Declaration, reveals a need for international support of technology in developing countries. One could argue that this support could encompass differential treatment of software composition in these countries to allow safe haven for native developers who accidentally infringe code on international software distributions. Otherwise, patent protection for software can discriminate against less powerful countries and contribute to the status quo in differential distribution of IP power.

The Balancing Act for Software Patents

Software patents may not support innovation. Yet as Samuelson (2007) argued, the dam has already burst; too much software located in powerful companies is now protected by patents for the courts or Congress to rescind patent protection for software. Burk and Lemley (2005) also see software patents as a now-permanent fixture in IP protection in the United States, and patent authority Robert Merges, author of an oft-cited 1999 critique of software patents, has more recently admitted that we are in a period of "normalization" of software patents (Merges, 2007)—a period in which they are simply part of the industry and not as destructive to its development as he had once argued. Allison and Hunter (2006) point out that even before the *State Street* decision in the United States, patents were issued to software simply because the lawyers had tied the software to a specific "machine" in their written description. Europe, despite a specific statement against the patenting of software *per se*, still issues patents for software (Blind et al., 2005; Nard et al., 2006). Therefore, patents on software are probably here to stay.

Given the inevitability of software patents, Burk and Lemley (2005) suggested that we should have a different approach to patent protection for software than we do for high-investment industries. They argue that a greater availability to more narrowly written patents might protect smaller firms from infringement and allow them to more easily acquire patents. Merges (1999) noted that we do not know for certain if software patents are bad or good for innovation, but we can all agree that bad patents are clearly bad; therefore, we should have higher scrutiny in business method patents, the category in which most software patents fall.

However, differential scrutiny of patents is simpler in theory than in practice. Allison and Hunter (2006) indicated that any special scrutiny of software patents will likely result in more circuitous written descriptions. They based this assertion on a review of a 2000 USPTO policy that implemented a "second pair of eyes review" (SPER) for business methods in an attempt to stanch the flood of these patents after *State Street*—a policy very similar to the one that Merges suggested. The USPTO reviewed only patents with business methods as a *primary* category, however, and so patent applications then spilled into related categories without a more thorough review (Allison & Hunter, 2006). As Allison and Hunter (2006) argued, definitional issues hamper special review; firms will circumvent the special regulations for software and simply file in a different category for a patent.

After the U.S. Supreme Court *Bilski* decision, it appears that any reform in U.S. patent law will need to come from Congress rather than the courts. Here is the key "passing the buck" passage in the U.S. Supreme Court's opinion:

> It is important to emphasize that the Court today is not commenting on the patentability of any particular invention, let alone holding that any of the above-mentioned technologies from the Information Age should or should not receive patent protection. This Age puts the possibility of innovation in the hands of more people and raises new difficulties for the patent law. With ever more people trying to innovate and thus seeking patent protections for their inventions, the patent law faces a great challenge in striking the balance between protecting inventors and not granting monopolies over procedures that others would discover by independent, creative application of general principles. Nothing in this opinion should be read to take a position on where that balance ought to be struck. (*Bilski v. Kappos*, 2010, pp. 9–10)

As the above passage suggests, patent law has become more difficult to understand at the same time that innovation is possible by a wider population. This theoretical balance in patent law to which the U.S. Supreme Court refers is struck differently across the globe, and it is perpetually inflected by international treaties, national legal codes, guidelines for patent examiners, and case law.

CONCLUSION

Patent protection for software can be a problem for writing researchers interested in the advancement of technology to support writing practices. Open-source alternatives to patented programs like Blackboard can be hamstrung by the intellectual property fences erected by companies flush with lawyers and other resources. Patent-supported monopolies can limit our freedom of expression and the means through which we distribute or teach writing. What are writing specialists to do? I argue that we should at least be aware of the relevance of patent law on the work we do.

Patent law is now part of the system of legal regulations for writing—the ways that writing is industrialized, constrained, and protected by law. Although patents were tested in the industrial economy and are often thought to be restricted to that economy, this is no longer true. Patents have come to affect the composition environments of software writers and technical writers for software—workers in the modern knowledge economy. As writers and writing researchers, our professional connection with computers draws us into the realm of patent law through the controversial and various patent protections for software across the globe. Writing studies can no longer bracket off patent law as irrelevant to our interests.

On a conceptual level, the application of patent law to software code—a kind of writing—should serve as a sign for us to reconceive our relationship to computers and the power of writing. Current applications of patent law can demonstrate just how fully integrated writing and coding are now; software can be thought of as a machine with physical actions in the world, but this specialized "machine" is a product of specialized writing. Patent law demonstrates how tricky it is to theorize software as a kind of writing *or* as a kind of technology; as a kind of expression *or* as a kind of action. It is all of these simultaneously, which opens up a deeper debate not only about how software and writing should be protected by IP law but also about how we should conceptualize the writing work we do.

REFERENCES

Allison, J. R., & Hunter, S. D. (2006). On the feasibility of improving patent quality one technology at a time: The case of business methods. *Berkeley Technology Law Journal, 21,* 729–794. Retrieved from http://www.btlj.org/data/articles/21_02_02.pdf

Baker, J. (2009, December 15). Blackboard, Desire2Learn announce patent cross license agreement and settlement of litigation. *Desire2Learn.* Retrieved from http://www.desire2learn.com/patent/

Bilski et al. v. Kappos. 561 U.S. (2010). *Slip opinion.* Retrieved from http://www.supremecourt.gov/opinions/09pdf/08-964.pdf

Bilton, N. (2010). An explosion of mobile patent lawsuits. *New York Times.* Retrieved from http://bits.blogs.nytimes.com/2010/03/04/an-explosion-of-mobile-patent-lawsuits/

Blackboard, Inc. v. Desire2Learn, Inc. 521 F.Supp.2d 575 (2007).

Blackboard, Inc. v. Desire2Learn, Inc. 574 F.3d 1371 (2009).

Blind, K., Edler, J., & Friedewald, M. (2005). *Software patents: Economic impacts and policy implications*. Cheltenham, UK: Edward Elgar.

Burk, D. L., & Lemley, M. A. (2005). Designing optimal software patents. In R. Hahn (Ed.), *Intellectual property rights in frontier industries* (pp. 81–108). Washington, DC: AEI-Brookings Joint Center for Regulatory Studies.

Brandir International, Inc. v. Cascade Pacific Lumber Co. 834 F.2nd 1142 (2d Cir 1987).

Brenner v. Manson. 383 U.S. 519 (1966).

Carmack, J. (1997, November 25). John Carmack—The Boot interview (with Alex St. John). *The John Carmack Archive.* Retrieved from http://www.team5150.com/~andrew/carmack/johnc_intervie_1997_John_Carmack_The_Boot Interview.html

Carnevale, D. (2007). Blackboard pledges to ease enforcement of controversial patent. *The Chronicle of Higher Education.* Retrieved from http://chronicle.com/article/Blackboard-Pledges-to-Ease/28490/

Diamond v. Diehr. 450 U.S. 175 (1981).

Essa, A. (2007). Software patents: Why should we care? *EDUCAUSE Review, 42*(2), 72–73. Retrieved from http://www.educause.edu/EDUCAUSE+Review/EDUCAUSEReviewMagazineVolume42/SoftwarePatentsWhyShouldWeCare/158127

European Patent Convention. (2000). *Article 52. Patentable inventions.* Retrieved from http://www.epo.org/law-practice/legal-texts/html/epc/2010/e/ar52.html

Feist Publications v. Rural Telephone Service. 499 U.S. 340 (1991).

Five IP Offices. (n.d.). *Five IP offices.* Retrieved from http://www.fiveipoffices.org/

Garfinkel, S., Stallman, R., & Kapor, M. (1991). Why patents are bad for software. *Issues in Science and Technology,* 50–55.

Gottschalk v. Benson. 409 U.S. 63 (1972).

Graham, S., Sichelman, T., Samuelson, P., & Merges, R. P. (2009). High technology entrepreneurs and the patent system: Results of the 2008 Berkeley Patent Survey. *Berkeley Technology Law Journal, 24,* 1255–1328. Retrieved from http://www.btlj.org/data/articles/24_feature.pdf

Hart, R., Holmes, P., & Reid, J. (2000). Study contract ETD/99/B5-3000/E/106: The economic impact of patentability of computer programs. *Intellectual Property Institute.* Retrieved from http://ec.europa.eu/internal_market/indprop/docs/comp/study_en.pdf

Heras, K. (2010). New Zealand to ban software patents. *The Next Web.* Retrieved from http://thenextweb.com/au/2010/07/17/new-zealand-to-ban-software-patents/

IBM's Supreme Court brief says that patents drive free software. (2009). *Slashdot.* Retrieved from http://yro.slashdot.org/story/09/09/03/1616220/IBMs-Supreme-Court-Brief-Says-That-Patents-Drive-Free-Software

In re Alappat. 33 F.3d 1526 (1994).

Jaffe, A. B., & Lerner, J. (2004). *Innovation and its discontents*. Princeton, NJ: Princeton University Press.

Johnson, L., Levine, A., & Smith, R. (2009). *The 2009 Horizon report.* Austin, TX: New Media Consortium.

Klemens, B. (2006). *Math you can't use*. Washington, DC: Brookings Institution.

Kuhn, B. (2010, August 16). *Considerations for FLOSS hackers about Oracle vs. Google* [Blog post]. Retrieved from http://ebb.org/bkuhn/blog/2010/08/16/oracle-google.html

Lenhart, A., Arafah, S., Smith, A., & Rankin Macgill, A. (2008). *Writing, technology and teens*. Washington, DC: Pew Internet and American Life Project.

Lessig, L. (2002). *The future of ideas: The fate of the commons in a connected world*. New York, NY: Vintage.

Lough v. Brunswick Corp. 86 F.3rd 1113 (1996).

Merges, R. P. (1999). As many as six impossible patents before breakfast: Property rights for business concepts and patent system reform. *Berkeley Technology Law Journal, 14*. Retrieved from http://www.law.berkeley.edu/journals/btlj/articles/vol14/Merges/html/reader.html

Merges, R. P. (2007). Software and patent scope: A report from the middle innings. *Texas Law Journal, 85*, 1627–1676.

Mueller, J. M. (2009). *Patent law* (3rd ed.). New York, NY: Aspen.

Nard, C. A., Barnes, D. W., & Madison, M. J. (2006). *The law of intellectual property*. New York, NY: Aspen.

Northern Telecom vs. Datapoint Corp. 908 F.2nd 931 (1990).

Park, J. (2005). Has patentable subject matter been expanded? A comparative study on software patent practices in the European Patent Office, the United States Patent and Trademark Office and the Japanese Patent Office. *International Journal of Law and Information Technology, 13*(3), 336–377.

Parker v. Flook. 437 U.S. 584 (1978).

Parloff, R. (2007). Microsoft takes on the free world: Microsoft claims software like Linux violates its patents. *CNNMoney.com/Fortune Magazine*. Retrieved from http://ec.europa.eu/internal_market/indprop/comp/index_en.htm

Patentability of computer-implemented inventions. (2005, July 6). *European Commission*. Retrieved from http://ec.europa.eu/internal_market/indprop/comp/index_en.htm

Samuelson, P. (1990). Benson revisited: The case against patent protection for algorithms and other computer program-related inventions. *Emory Law Journal, 39*, 1025–1154.

Samuelson, P. (2007). Software patents and the metaphysics of Section 271(f). *Communications of the ACM, 50*(6), 15–19.

Schonfeld, E. (2009). Here come the Twitter patent lawsuits. *TechCrunch*. Retrieved from http://www.techcrunch.com/2009/08/05/here-come-the-twitter-patent-lawsuits-techradium-?les-the-?rst-one/

Sengupta, A. (2004). The human right to development. *Oxford Development Studies, 32*(2), 179–203.

Shiva, V. (2001). *Patents: Myths and reality*. New Delhi, India: Penguin.

State Street Bank and Trust v. Signature Financial Group. 149 F.3d 1368, 1375 (1998).

USPTO and EPO work toward joint classification system. (2010, October 25). *European Patent Office*. Retrieved from http://www.epo.org/topics/news/2010/20101025.html

Vee, A. (2010). Carving up the commons: How software patents are impacting our digital environments. *Computers and Composition, 27*(3), 179–192.

Wikipedia. (n.d.). *History of virtual learning environments*. Retrieved from http://en.wikipedia.org/wiki/History_of_virtual_learning_environments

http://dx.doi.org/10.2190/LIIC10

CHAPTER 10

Orphan Works and the Global Interplay of Democracy, Copyright, and Access

Martin S. Copenhaver

Egalitarian access to information is a core ideal of a democratic society. Recognizing the fundamental right to participate in the ongoing cultural, social, and political dialogue, the framers of the U.S. Constitution ensured this access by providing that "Congress shall have the Power . . . To promote the Progress of Science and useful Arts, by securing for limited Times to authors and Inventors the exclusive Right to their respective Writings and Discoveries" (Art. 1, Sec. 8, Cl. 8, hereinafter the "copyright clause"). This notion of egalitarian access to information is crucially important because access facilitates the exchange of ideas through an ongoing dialogue that enables democracy itself (Herrington, 2011).

As both users and creators of copyrighted materials, technical communicators experience this importance of access to information, copyrighted or otherwise, in their day-to-day activities. On the one hand, technical communicators in academia use this access to copyrighted works to comment on and critique various communicative structures. On the other hand, technical communicators working in industry access these materials in order to create new communication products (e.g., user manuals and online help systems) used by a range of clients or consumers. In both cases, technical communicators must obtain the appropriate permissions to use copyrighted works to create new materials should the intended uses exceed the allowable exceptions to or limitations of copyright. The rise of digital media, however, has created new contexts that can hinder these processes, and one such situation involves orphan works, or works for which the copyright holder cannot be identified and located.

The purpose of this chapter is to uncover how orphan works affect technical communicators in their daily interactions on a global stage. To satisfy this objective, the chapter first provides an overview of orphan works. Next, the author examines the harms to innovation, speech, and democratic interactions that can arise from the orphan works situation. The author then looks at the global nature of orphan works and considers the notions of control of law and "harmonization" as they relate to the orphan works problem. The chapter concludes with a discussion of potential best practices technical communicators can use to address the situation of orphan works. Such an examination of these issues can help technical communicators more effectively and proactively approach the issue of orphan works in relation to their work.

ORPHAN WORKS: A BRIEF OVERVIEW

While there is disagreement over the precise definition of an orphan work, as is apparent in the work of the U.S. Copyright Office in its foundational report on the matter, it is generally accepted that an orphan work is a copyrighted work for which the copyright holder "cannot be identified and located by someone who wishes to make use of the work in a manner that requires permission of the copyright owner" (U.S. Register of Copyrights, 2006, p. 1).[1] Such works typically arise in the "rights clearance" process through which a potential user who wishes to make use of a work's copyright beyond an allowable exception seeks out the copyright holder for permission (typically in the form of a licensing agreement) to use the copyright. Orphan works pose particular challenges in a digital age wherein society increases its reliance on materials that would traditionally be found in physical libraries (Proskine, 2006).

Because the copyright holders of certain works contained in libraries and archives cannot be located, these libraries and archives avoid continuing with the infringing digitization of these materials for broad access. This avoidance is based on the fact that using such works without the permission of the copyright holder could result in litigation should the copyright holder appear after the infringing use has begun. A specific example of how orphan works arise in practice for libraries is illustrated by the Library of Congress' attempt to digitize letters, magazines, and other materials relating to Hannah Arendt (Hughes &

[1] Following the U.S. Supreme Court decision in the case of *Eldred v. Ashcroft* (2003), the U.S. Copyright Office undertook a large effort in 2005 in order to understand the scope of the problem of copyrighted works for which identifying the copyright holder is difficult or impossible (a concern brought up in Justice Breyer's dissent in the case). Soliciting input from libraries, archives, authors, and many others, the Copyright Office received over 850 initial and reply comments. In 2006, the Copyright Office synthesized these responses in its *Report on Orphan Works,* leading to a better understanding of the definition and scope of the orphan works problem, its legal context, and possible solutions.

Cohen Harrington, 2005). Using numerous databases such as telephone directories, websites, and various library catalogs, several Library employees spent 2 years attempting to determine copyright ownership of nearly 7,000 items (out of a total of 25,000 items, the majority for which the Library abandoned copyright holder inquiries due to the potential cost) (Hughes & Cohen Harrington, 2005, p. 6). Throughout the process, employees encountered many problems, such as for works of corporate authorship for which copyright ownership was particularly difficult to determine given mergers and name changes; even for those works for which a copyright holder was identified, it became quite difficult to actually locate the copyright holders in order to seek permission to digitize the materials. In the end, fewer than 500 copyright holders responded to a permission request (out of 1,932 inquiries sent); as described bluntly by the Library, the project was "severely hampered by 'orphan works'" and likely would not have been pursued at all if it had not been for "generous outside funding" (Hughes & Cohen Harrington, 2005, pp. 6–7).

Because most libraries and archives avoid assuming the risk of infringement, access to these more "traditional" forms of orphan works is, at best, limited or, at worst, nonexistent. Researchers who have focused on this situation (e.g., Huang, 2006; Proskine, 2006) consider this approach paradoxical and suggest it is fundamentally at odds with the promise of novel technologies to allow for broader, possibly global access to copyrighted works. In fact, this situation ultimately threatens the very underpinnings of the U.S. copyright clause, for copyright subsists in works merely as a means of a achieving the more important ends of learning through regulated knowledge distribution (Patterson, 1987). Consequentially, when society cannot access copyrighted works, the regulatory structure of statutory copyright fails to achieve its constitutionally assured purpose.

Yet orphan works are not limited to the works contained in the depths of libraries and archives. Another type of orphan work includes the various ephemera on the Internet, and such works include weblogs, numerous websites, and other communicative structures. These ephemeral forms of copyrighted works have an increased likelihood of being "orphaned," particularly because of the mere quantity of works online and the potential difficulty of identifying and locating their respective copyright holders in an environment in which identity can be very difficult to discern (Huang, 2006; Hughes & Cohen Harrington, 2005). Furthermore, it is important to note that these ephemera expose a different problem than orphan works contained in libraries and archives, which threaten the ability of society at-large to access information. Instead, as Huang (2006) suggests, orphaned ephemeral works make innovation and creativity much more difficult, particularly for those whose work builds upon existing copyrighted expressions.

This situation is particularly true for technical communicators who rely on these ephemeral documents in their work—academicians in assessing and critiquing these various forms of communication and practitioners in developing more effective and innovative forms, models, or structures. While technical communicators might still be able to access these potentially orphaned works for their ideas, any intended use of the copyrights associated with these works (beyond an allowable exception to or limitation of copyright) would entail a search that researchers suggest is likely to fail (Huang, 2006). Thus, technical communicators should have an interest not only in the effects of the more traditional orphan works on the broader notion of access, but also the effects of orphan works, particularly ephemera, on the ability of technical communicators to critique, analyze, and develop effective, novel communicative forms.[2]

For example, an American technical communicator may be interested in adapting an existing copyrighted communicative structure she has found on the Internet in order to create a boilerplate for a client. If her desired use of the structure goes beyond the allowable exceptions to or limitations of copyright, she would need to seek out the copyright holder for permission; however, the copyright holder could potentially be difficult to identify and locate, particularly given the global, ephemeral nature of the Internet. Should a diligent search for the copyright holder fail, in which case the work could be considered orphaned, she would either be unable to use the structure or risk infringement if she proceeds with her intended use. As this example demonstrates, orphan works diminish the ability of technical communicators to innovate effectively and efficiently.

This issue of orphan works becomes even more important for technical communicators when it is placed within a global context. To understand the orphan works situation from this perspective, it is necessary to examine the relevant treaties and multilateral trade agreements that currently control the international intellectual property (IP) "market" for copyrighted works. Through this international IP structure, the importance of questions regarding the harmonization of solutions and control of law becomes apparent. Particularly important is the global nature of the Internet. Within a global setting, very little is agreed upon in the international legal community about which laws are relevant in cases of infringement when the infringing works were disseminated across the Internet and thus potentially into hundreds of different nation-states with differing treatments of intellectual property (Wilson, 2009).

These factors become even more problematic in the context of orphan works, because the existing nation-specific solution structures are unique and address the

[2] A troubling point that is beyond the scope of this chapter regards these ephemeral documents once they are removed from the Internet. For all intents and purposes, they are inaccessible, yet they still have an associated copyright. This in itself warrants a thoughtful analysis, particularly in light of the purpose of the copyright in the United States. Technical communicators would benefit greatly from access to the ephemeral documents that were once on the Internet and have subsequently been removed; further research should suggest ways of addressing this problem.

issues of search requirements and infringements differently, if at all. While there are no answers to the control-of-law quandary, it provides motivation for technical communicators, whose work environments are becoming increasingly global, to understand harmonization, control of law, and how they play a role in the orphan works situation. Moreover, the global nature of the interactions between technical communicators and their colleagues as well as their audience presents the possibility for technical communicators to shape the direction of the dialogue surrounding orphan works. Therefore, it is important for technical communicators, even those working exclusively within domestic environments, to understand the potential effects of their interactions with orphan works on both a local and a global stage. In the following sections of this chapter, I explore in depth these notions that arise in the global context. First, though, it is important to consider how orphan works affect broader notions of access, challenging the constitutional foundations of copyright; this ultimately affects the ability of technical communicators to participate in democratic interactions.

WHY ORPHAN WORKS MATTER

Speech and Innovation

Speech and innovation are central to the work of technical communicators. Academic technical communicators have speech interests in being able to analyze, critique, and comment on the possible impact of varying forms of (copyrighted) communication. At the same time, practitioners employ existing structures and work to develop innovative, potentially more effective products, structures, and documents. Yet speech is not an interest unique to academics, and likewise, innovation is not solely of interest to practitioners. For many practitioners working within a work-for-hire environment, their ability to "speak" is correlated with their ability to create products that "have a broad effect on the general public's access to shared communication over a broad range of topics" (Herrington, 2011, p. 67). Likewise, academics who analyze various structures have an interest in developing more effective forms of communication by understanding the problems with existing forms. It follows that technical communicators should have an interest in how orphan works damage their abilities to work effectively in environments structured around interests in innovation and speech.

To begin, the effects of orphan works on innovation are particularly concerning. Consider a technical communicator in industry designing an online user interface for a client company. Suppose that while developing the user interface, he or she comes across an existing copyrighted interface that he or she would like to build upon to make a more effective structure for her own client. In the process of acquiring the appropriate permission from the copyright holder in order to make a derivative work, that technical communicator would need to locate and identify the copyright holder. If the copyright holder cannot be located (which is a possibility, given the quantity of creative works online), the technical communicator would be unable to continue with his or her intended use or fear infringement should the copyright holder reappear after the use has begun (Huang, 2006). This situation ultimately stifles the technical communicator's ability to innovate, for innovation requires access to copyrighted materials as well as the appropriate permissions. Even though this technical communicator could consider alternative structures for possible innovation, the situation is often one wherein alternative sources from which creative, innovative expressions can be developed will not suffice (Van Houweling, 2005).

Innovation for technical communicators thus becomes much more difficult in light of orphan works. The increased likelihood of searches for copyright holders of ephemeral works to fail can discourage potential innovators from pursuing permission for use of any works, much like the many libraries and archives that abandon all searches before beginning due to the potential pitfalls ahead (Hughes & Cohen Harrington, 2005; Nisbet, 2005). This potential stifling of creative efforts should be considered particularly disconcerting, especially for technical communicators whose work requires an interaction with ephemeral documents for which orphaning is a valid concern.

However, as the work of academic technical communicators throughout the past two decades has demonstrated, exceptions to or limitations of copyright (such as fair use in the United States) are viable avenues of pursuit for technical communicators, particularly for those whose work merits speech (Herrington, 2011). Even for technical communicators working under these exceptions and limitations (for which permission of the copyright holder is unnecessary, in certain circumstances), orphan works are still relevant.

Support for this position is found in the global nature of the work of technical communicators. Oftentimes technical communicators, particularly those developing products, structures, or documents for online consumption, work for clients throughout the world. In their daily work, these technical communicators strive to effectively convey information to their target audience—a situation that inherently relies on understanding that audience. For an American technical communicator producing user help manuals for a client in France, it would be important to learn more about the French audience. To do so, it would be imperative to have access to existing French communicative structures, not only those found on the Internet. However, the ability to access copyrighted materials is the crux of the orphan works problem, particularly when it comes to those works located in the depths of libraries and archives. When many of these orphan works are unavailable for technical communicators to use to develop effective materials, technical communicators are hampered in their daily interactions. It follows that orphan works affect all technical communicators, even those who rely on the exceptions to or limitations of copyright in carrying out their work.

Furthermore, there are also concerns about the effects of orphan works on speech. For all technical communicators, there are interests in critiquing and commenting on various existing structures, documents, and forms in order to participate in a constructive dialogue and to guide progress and developments in the field. Much as orphan works damage innovative efforts and creative work practices, orphan works, particularly those for which access is limited, diminish the ability of technical communicators to make commentary. For those in industry, this commentary is primarily in the form of innovation and novel, creative developments, so the findings above regarding innovation apply. For technical communicators in academia, speech depends upon access to copyrighted works. While there may be access to many ephemeral works online, research suggests that access to orphan works contained in archives and libraries can be incredibly limited, greatly diminishing the ability of technical communicators to make effective critical commentary (Huang, 2006).

Though beyond the scope of this chapter, it is important to note a particularly disconcerting aspect of potentially orphaned ephemeral works: the fluid alteration of content facilitated by the Internet can allow copyrighted works to be "lost" from the Internet, after which all access is impossible. However, being able to analyze why certain content or structural changes were made would necessarily be a point of consideration for academicians who assess how dynamic choices affect efficacy and for practitioners who want to better understand why certain choices in content development and management are more effective than others. Even though the potential loss of these ephemeral works is not explicitly connected to orphan works, it raises similar concerns about the ability of technical communicators to work effectively and efficiently in global, Internet-driven environments.

Admittedly, many technical communicators work in environments in which practices such as single sourcing and the use of boilerplates are common and in which innovation and speech interests are, out of necessity, not the focus of their work. Yet even those working in such environments are not immune to the notions of speech and innovation, and are thus affected by orphan works. This is for several reasons. First, the employers who create settings that encourage consistent document production nevertheless are interested in the efficacy of the documents being created; the technical communicators producing these materials, then, have interests in more effective and innovative techniques being developed by other technical communicators.

Furthermore, and perhaps more importantly, the work of technical communicators is unique in that it has a broad effect on society, so even those working in limiting environments have the potential to have an impact on society through their work (Herrington, 2011). Their ability to shape the perception of information suggests that technical communicators, regardless of the limitations of their environments, would have an interest in their speech and innovative powers implicated in their works (Herrington, 2011, p. 67). Finally, for those who work solely in environments wherein the impact of their work on society is not directly implicated, such as those who construct technical reports for internal corporate consumption, there are broader implications in democratic interaction that are important to consider. This situation serves as the focus of the following section, wherein I primarily consider the orphan works that fill libraries and archives, for which public access is limited at best.

Democratic Interactions

At the end of the day, technical communicators are participants in a democratic dialogue. As such, they are affected by the broader interests satisfied by access to knowledge, such as democracy and free speech. These interests are present throughout the U.S. Constitution, particularly in the copyright clause. Because the goal of copyright is to promote the dissemination of knowledge for the purpose of learning and the general advancement of society, it becomes clear why access to knowledge in copyrighted materials is inherently connected to the broader ideals of a democracy (see e.g., Herrington, 2001; Patterson, 1987).

Herrington (2011), for example, interprets this connection by considering the intent of the framers. She suggests, and I strongly agree, that "democracy, free speech, self-actualization, and humanistic endeavors stem from this core [inherent in the copyright clause] in a mutually dependent interplay of support for the [United States'] principles" (p. 49). This interpretation that "progress" means the advancement of knowledge and learning and thus supports broader, democratic notions is supported through careful consideration of the historical developments in copyright in the United States throughout the past three centuries, as the influential work of copyright scholar L. Ray Patterson (1987) makes clear.

Consequently, when the public cannot access the ideas (which are not copyrighted) or the expressions (which are copyrighted) within copyrighted works, society cannot "progress." This does not affect only our ability to learn, it affects our ability to participate in a dialogue. As a dialogue necessarily entails mutual criticism and commentary, it cannot go on if there is nothing upon which to comment. These theoretical positions may seem overstated; yet consider an orphan work "buried" in the depths of a library. Statistics produced by the Library Copyright Alliance (Nisbet, 2005) would suggest that, if the work were older or of a nontextual form, or worse, unpublished, a search for the copyright holder would likely prove unsuccessful, in which case the library would likely not be willing to provide online access to it, even for a select group of researchers. In this sense, the ideas contained within the work, despite being theoretically, in some sense, a part of the "public domain," are confined to the physical library, where only those with physical access are able to benefit from the work's contents (Patterson, 1987, p. 7).

This concept that ideas are limited to the physical domains of a library presents a paradox of sorts, for the author who incorporated public domain ideas into her work is "withholding," not through an assertion of her own exclusive (copy)rights in the work, but merely because she cannot be identified and located, the ideas as well as her original expressions in the work from being redistributed back to the public for consumption. Yet "the public can be compensated most effectively" for an author's "encroachment on the public domain . . . by making the author's efforts accessible" (Patterson, 1987, p. 7). With orphan works, though, such redistribution is impossible unless the entity wishing to provide broad access is willing to risk a claim of infringement should the copyright holder reappear after the infringing use of the copyright has begun.

This example, which anecdotal evidence suggests is relatively commonplace, supports the conclusion that orphan works, particularly those in libraries and archives, portend an ideological conflict in copyright. When access to copyrighted works is not sustained, knowledge is not disseminated; when knowledge is not disseminated, a dialogue cannot take place; when a dialogue cannot take place, the more fundamental notions of self-actualization and free speech are threatened (Herrington, 2001). These concerns suggest that orphan works impact the ability of technical communicators, as members of society, to participate in a democratic system. Thus, technical communicators should care because orphan works hamper democratic dialogue and the dissemination of knowledge for the mutual benefit of society, even if technical communicators can access certain potentially orphaned ephemeral documents. In this sense, though, everyone should care about orphan works because of their potential limitations on the ability of society to shape the development of culture.

Technical communicators work in many different environments throughout many different cultures; moreover, they complete an incredibly varied body of work, from technical reports to creative communicative products. All technical communicators, then, have varied interests that are a result of their cultural and environmental influences, among others. Yet at the core of the work of all technical communicators are interests in speech and innovation. These interests enable technical communicators to maintain an active role as participants in a democratic dialogue through the creation of many different products, structures, and critiques that ultimately serve to affect the way society perceives content. As a result, when orphan works threaten the fundamental notion of access to copyrighted works, technical communicators should care. Not only do orphan works have very practical effects on the ability of technical communicators to work effectively, they also threaten broader humanistic interests, such as the notion of egalitarian access to information. And when society does not have access to copyrighted information, the democratic ideals of free speech and self-actualization break down.

Therefore, the consideration of the orphan works situation within the limitations of the U.S. intellectual property foundations is important for understanding the detriments to the constitutionally assured notion of access to copyrighted information that allows for the purpose of copyright to be fulfilled: the promotion of learning through the dissemination of knowledge. Understanding the orphan works problem from this perspective is important for technical communicators, who have significant interests in innovation and speech that are supported by the U.S. Constitution.

The interactions of technical communicators, however, take place on an increasingly global stage. As such, technical communicators should have an interest in orphan works within a global context wherein various ideological conflicts arise and affect how orphan works are treated transnationally. The developments in this global context will motivate several suggestions for how technical communicators can effectively and proactively participate in the ongoing orphan works dialogue through practices for both academics and practitioners. First, though, the global context is explored, particularly how ideological conflicts arise and how the elements of the international intellectual property framework affect future developments in orphan works. Then, consideration of control of law and the notion of harmonization of solutions will shed light on the best practices for technical communicators to carry out in their ongoing work.

THE GLOBAL NATURE OF ORPHAN WORKS: SOLVING THE PROBLEM

Ideological Conflicts

The work of technical communicators as content managers, product developers, commentators, and innovators is becoming global in nature. Many practitioners, for example, complete work for clients across the globe or work in environments such as globally distributed virtual teams. Likewise, many academicians access, critique, and comment on structures on the Internet. In fact, the very nature of the Internet diminishes the traditional notion of borders and creates an environment in which people on opposite sides of the globe can communicate quickly and easily.

At the same time, orphan works are not unique to the United States. Several nations, including Canada, Japan, and the United Kingdom, already have orphan works solutions implemented in the form of national legislation that aims to alleviate the orphan works problem. Yet all existing national solutions are unique, and there does not exist a global solution to the orphan works problem (Wilson, 2009). As a result, technical communicators who access potentially orphaned, online ephemera should have an interest in how inconsistent solutions would affect their ability to effectively address the problem in their own work. At the same time, it is important to understand how the structure of the current international intellectual property framework could potentially affect developments in orphan works solutions, both locally (at the national level) and globally.

This international intellectual property framework is composed primarily of treaties and trade agreements. These include the Berne Convention for the Protection of Literary and Artistic Works (1971) (Berne Convention or Berne) administered by the World Intellectual Property Organization (WIPO) and the Agreement on Trade-Related Aspects of Intellectual Property Rights (1996) (TRIPS Agreement or TRIPS) administered by the World Trade Organization (WTO), among others.[3] The Berne Convention sets certain statutory minima regarding the protection of intellectual property rights necessary for all nations that are members to the Convention. The TRIPS Agreement incorporates and expands on the rights established in Berne (except for certain *droit moral*, or moral rights), and its inclusion as an agreement administered by the WTO requires all WTO members to comply or face possible trade sanctions. Many nations are parties to both Berne and TRIPS, and so they serve as the defining framework for all developments in intellectual property across the globe.

However, this international IP structure has foundations that are very different than those of the United States. Whereas U.S. intellectual property law is derived from the constitutional policy of ensuring the promotion of learning, TRIPS and Berne are founded on the principles of ensuring "maximum [economic] value for intellectual property products in the global market" (Okediji, 2000, p. 81). These governing documents, particularly TRIPS, focus on the "markets" for intellectual products and are mostly built on the principles of free trade (Birnhack, 2006). These ideological differences between trade and access are very relevant for many aspects of copyright, but particularly for orphan works. As the instructive example of the U.S. orphan works situation demonstrates, orphan works pose very real threats to knowledge and democracy itself. Placing the orphan works problem in a trade context reduces it to purely economic terms in which the economic inefficiencies of the permissions clearance process likely serve as the main motivation for resolving the orphan works problem.

Example Situation: The Case of Fair Use

In order to understand the ideological difficulties of extrapolating from the access-driven basis of U.S. intellectual property law to the trade-driven basis of international IP law, it is important to consider the ongoing dialogue about the creation of an international fair use doctrine. The fair use doctrine, as statutorily codified in the United States with the 1976 Copyright Act (specifically, 17 U.S.C. Sec. 107), allows for some of the policy goals of the U.S. Constitution,

[3] Other such documents that comprise the international intellectual property framework are the WIPO Copyright Treaty and the WIPO Performances and Phonograms Treaty. These texts build upon Berne Convention and TRIPS to address technological issues, and so solely considering the Berne and TRIPS shall suffice for the present purposes. For a broader view of the international intellectual property structure, see Birnhack (2006).

such as access to copyrighted material for learning and dialogue, to be effectuated (Patterson, 1987).

The core of the fair use doctrine rests on allowing for exceptions to copyright in certain instances of speech and innovation. This approach is consistent with the intentions of the framers of the Constitution in supporting egalitarian access to knowledge for a thoughtful, critical dialogue that results in the development of knowledge for the benefit of society. While not explicitly obvious, the ideals of fair use are consistent with the notions hindered by orphan works: speech, innovation, and democratic interactions. It follows that fair use is incredibly relevant to the orphan works problem, not because it applies directly but because thinking about orphan works in terms of speech and innovation suggests a fair use structure through which innovation and speech are supported.

And so, when problems arise with creating an international structure for fair use within the trade-driven IP framework, ideological problems arise in considering orphan works. Particularly difficult in the case of creating an international fair use structure, which still does not exist, is the connection between the fair use doctrine and the goals of the U.S. Constitution, particularly the copyright clause and the First Amendment (Herrington, 2010; Okediji, 2000). To effectively construct an international fair use framework, it is necessary to rely on advocating for a structure that supports the deeper concepts inherent in fair use. This kind of egalitarian access to information helps support goals of self-actualization and democratic interaction, for it is impossible for the U.S. foundations to placate other intellectual property structures by being incorporated into TRIPS or Berne (Herrington, 2010). However, as Herrington (2010) aptly recognizes, there still arise problems in simply advocating for a conceptual structure. This is because its creation necessarily relies on ensuring that less powerful nations who are affected by the international IP framework are not marginalized in ongoing discussions. At the same time, a fair use structure, as it is tied to the U.S. Constitution, supports broader notions of human rights, and the lack of global consensus on human rights alone makes developing a global structure difficult (pp. 324–327).

Therefore, taking the orphan works situation and placing it in the global orphan works context implicates many ideological conflicts that could arise. An international fair use structure would be crucially important for technical communicators to function effectively in the global workplace (Herrington, 2010). At the same time, the ideological connections between orphan works and fair use suggest ways in which technical communicators could take place in ongoing dialogues about orphan works solutions. More importantly, though, as the international fair use developments make clear, there arise important humanistic considerations that affect technical communicators. These humanistic considerations are further explored later in this chapter.

First, though, it is necessary to consider how some of the transnational, ideological conflicts manifest themselves as actual problems in working toward a global solution. To understand these differing approaches to solutions, I briefly consider the two main approaches to solving the orphan works problem locally (at the nation level). I then use the global level to discuss how these solutions fit into the international IP framework and how the IP framework affects the development of solutions. The notion of control of law is further considered, as well as how developments in this area could affect technical communicators. This leads to a consideration of the broader impact of motivations toward harmonization on technical communicators.

Conflicts in Solutions: When Ideological Turns to Actual

There are two primary approaches to solving the orphan works problem locally: a limitation on remedies approach and a compulsory license approach.[4] The limitation on remedies approach has been the primary consideration in potential U.S. legislation. It focuses on reducing the liability for infringers of orphan works by limiting the monetary and injunctive relief available to reappearing copyright holders of (once-)orphan works. Additionally, this approach applies primarily to users who have made use of the copyright associated with an orphan work after having conducted a diligent search for the copyright holder and failed to successfully locate and identify him or her (U.S. Register of Copyrights, 2006). This approach also allows for users to make use of the copyright without verification that a search was "diligent," and so should a copyright holder reappear after the infringing use has begun, the diligence of the search would need to be addressed on an *ad hoc*, case-by-case basis (Thompson, 2006).

In contrast to this approach is the compulsory license approach generally adopted in various forms in Japan, Canada, and South Korea, among others. In these compulsory licensing systems, a public body typically assesses the diligence of a failed search for the copyright holder conducted by a potential user of the orphan work's copyright (Wilson, 2009). Based on varying search standards, these public bodies might issue a compulsory license for certain uses of the copyright if the search meets certain minima. This solution is drastically different from the limitation of remedies solution, which takes an *ad hoc* approach to determining the validity of a diligent search. Moreover, the differences between the two unique structures raise very different legal issues, which, despite being beyond the scope of this chapter, would be necessary to consider fully if movements toward harmonization gain force.

[4] There is another approach, extended collective licenses, which is employed by some Nordic countries (Wilson, 2009). However, for the present purposes, the two primary solutions that have been considered will suffice since these simply serve to show the conflicts between local (national) solutions.

While these solutions are local, the global nature of the international IP framework directly affects their development. Berne and TRIPS do not contain standards for local orphan works legislation. They do, however, limit exceptions and limitations to the exclusive rights of the copyright holder (Ginsburg, 2008). For example, the Berne Convention (1971) requires that exceptions to the exclusive right of reproduction be limited to "[a] certain special cases, provided that such reproduction [b] does not conflict with a normal exploitation of the work and [c] does not unreasonably prejudice the legitimate interests of the author" (Art. 9[2]). This test, generally known as the "three-step test," was expanded to all exclusive rights of the copyright holders by the TRIPS Agreement (Art. 13). The relationship between Berne and TRIPS is still debated among scholars of copyright, especially given a WTO Dispute Settlement Panel's (2000) interpretation of the three-step test in proceedings regarding 17 U.S.C. Sec. 110(5) governing exceptions to copyright for a subset of public performances.[5] Regardless of the interpretation of this relationship, though, it is important to recognize that national laws are limited in their abilities to provide exceptions to the exclusive rights of creators of intellectual products. Huang (2006) suggests that these limitations ultimately make the construction of feasible legislation that would pass international muster difficult. As Ginsburg (2008) and Thompson (2006) demonstrate, making any orphan works legislation fit within the narrow confines of the international IP framework requires very thoughtful, careful consideration.

Control of Law

The local, nation-specific approaches to the orphan works problem and the limitations on local solutions should be considered important to technical communicators because a patchwork of only partially effective solutions will make functioning within a global environment incredibly difficult. An important and practical question for technical communicators is, "Which country's laws govern how orphan works are treated, especially in situations in which work is transmitted via the Internet wherein potential infringement can be taking place in many locations simultaneously?" For a practitioner working for U.S. organizations that single-sources material to France, understanding this control-of-law question is important for determining how to proceed in approaching potentially orphaned works. For technical communicators who access online materials

[5] A strict interpretation suggests that TRIPS's expansion of the three-step test applies only to new rights set out in TRIPS, and thus Berne's minor exceptions doctrine focusing on *de minimis* exceptions would be applicable to any proposed exceptions to rights set out originally in Berne. Another prevailing interpretation is that TRIPS applies the three-step test to all rights set out in Berne, and thus the three-step test should be applicable to proposed exceptions. See WTO (2000), Thompson (2006), and Ginsburg and Goldstein (2005).

that are potentially orphaned works originating from other countries, the control-of-law question affects which orphan works legislation would govern his or her actions regarding searches and the associated, necessary requirements for the copyright holder. In the end, control of law as it relates to the international IP framework affects all technical communicators in almost all situations, regardless of whether the works are orphaned, because control of law affects how all copyrighted works are treated transnationally (Wilson, 2009).

Yet control of law in the international intellectual property framework is currently out-of-date, especially as it pertains to questions of infringement on the Internet (Wilson, 2009). The primary control-of-law clause is in the Berne Convention, and this clause states that "the extent of protection, as well as the means of redress afforded to the author to protect his rights, shall be governed exclusively by the laws of the country where protection is claimed" (Art. 5[2]). Whereas in traditional cases of infringements this "exclusivity" requirement is reasonable, the country of infringement is much harder to determine in an environment in which copyrighted works are disseminated cheaply and easily. Thus, infringement is possibly widespread. Consider a hypothetical situation in which a person in the United States puts illegally copied manuscripts (of Canadian authorship) onto a server (located in the United States), and these manuscripts are accessed by users in Canada. In this situation, which country's laws should apply to infringement by the person illegally providing copies?

A similar question was raised in the French court case of *Société des auteurs des arts visuels et de l'image fixe (SAIF) v. SARL Google France* (2008) (SAIF). In this case, the court addressed a challenge to Google's Image Search for violation of the French copyright holders' exclusive rights to reproduce their works. The court chose to apply U.S. law, as the United States was the location of Google's servers where the literal infringement (the accessing of the servers where Google's Image Search linked to the French copyright holders' images) occurred. Wilson (2009) suggests that the SAIF court's decision to apply the United States as the place of causation instead of French law, as the place of effect has potential harms for future developments in this area. Under the SAIF reasoning, a savvy infringer could place servers in a country with minimal IP protections and be free from infringement claims.

As this case and others demonstrate, the control-of-law clause in the Berne Convention provides little guidance for courts in cases in which infringement takes place across the globe, especially via the Internet (Wilson, 2009). This factor is important to orphan works, particularly as they pertain to technical communicators, for control of law necessarily dictates which orphan works solution applies in which instances. Currently, there are no solutions that rectify this control-of-law quandary. As long as this problem persists, the ability of solutions to the orphan works problem to take hold on a global scale is limited. If, for instance, future developments in Berne's control-of-law clause require that the search requirements for an orphan work be consistent with the work's country of origin, it would likely still require a great deal of funds to simply identify the country of origin in order for a potential user to even know which requirements apply (Ginsburg, 2008). While the possible solutions to this problem are beyond the scope of this chapter, it is important to recognize that technical communicators should have an interest in which laws apply not only for cases of orphan works but for all works. All technical communicators produce works that are affected by control of law and any possible future developments in it. Thus, technical communicators should maintain an interest in the developments in this area of international IP law as they take place throughout the next decade as the Internet further challenges the traditional notions of physicality upon which Berne is based.

"HARMONIZATION?": ORPHAN WORKS AND HUMANISTIC CONCERNS

The notion of harmonization is significant throughout all of intellectual property law. After all, consistency across borders served as some of the primary motivations for the Berne Convention and the TRIPS Agreement (Herrington, 2010). Consistency allows for more efficiency and reduces the transaction costs of working in a global environment. A globally consistent solution would likely improve the ability of technical communicators to conduct business efficiently and effectively. For example, those technical communicators working with single-sourcing would likely benefit from greater access to copyrighted materials from the environments to which they are distributing content. Increased access would allow these practitioners to make more effective decisions about how to convey content in a means consistent with that environment or community's expectations. Even technical communicators who function within a single domestic environment would experience positive effects from a harmonized orphan works solution. This is because harmonization would likely increase global access to information, thus increasing the strength of the democratic process and furthering the development of knowledge.

However, the overriding rhetoric regarding orphan works and harmonization in general should be assessed carefully. As with the creation of an international fair use doctrine, there are broader humanistic implications that are important to consider. Any global solution would likely set forth precise guidelines for search requirements that rely on the use of electronic databases. Imposing requirements for the construction of databases on developing countries or least-developed countries (LDCs) could be incredibly problematic.

As has been demonstrated in the push to construct an international fair use doctrine, developing countries lack an ideological foundation for intellectual property law. Moreover, many of these nations are bound by TRIPS simply because they need to be members of WTO in order to successfully trade in all areas of commerce (Birnhack, 2006). Yet the orphan works problem arises through changing attitudes about technology, access, and copyright. Developing countries have neither a strong IP foundation nor technological capabilities. As a result, they are unlikely to be positively affected by orphan works developments. In fact, of great concern to technical communicators should be that a global orphan works solution incorporated into TRIPS would likely devastate the body of intellectual products in developing countries and LDCs.

This devastating power can be seen in how an orphan works solution, whether focused on a limitation of remedies or some form of a licensing system, would function. Regardless of the form, a system would in some way prescribe the necessary conditions for conducting a diligent search for the copyright holder. Given the global nature of the problem, these searches would likely entail use of the Internet.

While developed countries have at their disposal resources to create these databases, developing and least-developed countries do not. Any "search" for most copyright holders in these countries would likely prove unsuccessful given the lack of technological resources that could possibility be devoted to the orphan works problem. This situation opens the door for exploiters of the system to severely damage the cultural heritage contained within the copyrighted works of these countries by conducting "searches" (which the exploiter likely knows will be unsuccessful) in order to declare the vast amounts of cultural works "orphaned" and thus, consistent with the means of going forward dictated in the solution, may proceed in using the copyrights. This abuse of the system may seem extreme, but its potential of occurring is increased in light of recent abuses of culturally valuable works from developing countries (Herrington, 2010).

Even though it may be possible to develop a globally consistent orphan works solution to the mutual benefit of creators and users, the potential marginalization of developing and least-developed countries should be of concern to technical communicators who carry out their work with humanistic interests in mind. While the speech and innovation interests of technical communicators may be furthered by a consistent solution, I believe that technical communicators, whose work is unique in that it affects how society perceives content, should thoroughly consider the wisdom of imposing an orphan works solution on countries that are not prepared (by ideological foundation or by technology) to address the orphan works solution at this time.

BEST PRACTICES FOR TECHNICAL COMMUNICATORS

As St.Amant and Rife (2010) suggest, cross-cultural "friction points" arise through the work of technical communicators and should be addressed by technical communicators who work at the center of the global exchange of content (p. 249). The orphan works problem, which arises in the friction point of copyright, affects how technical communicators speak and innovate in their daily practices. Therefore, all technical communicators should have an interest in participating in the ongoing dialogue about orphan works and how they should be treated on a global scale. Through their work, technical communicators should keep in mind the negative impacts that orphan works have on the broader notions of access, democratic interactions, and self-actualization. What follows are strategies technical communicators can use to approach orphan works effectively and proactively:

Strategy 1: Participate in Ongoing Discussions About Types of Works That Should be Addressed with Orphan Works Legislation

Some technical communicators work in environments such as medical communication that require an understanding of how the notions of privacy and disclosure vary across borders. Applying their understanding about constructs of privacy, these technical communicators could help to address the unanswered questions about unpublished works and whether these should be included in orphan works legislation, or whether the concerns for privacy and the notion of the moral right of *droit de divulgation* should take precedence.

Strategy 2: Contribute to the Development of Effective Databases for Copyright Holder Inquiries

In their work, technical communicators evaluate the efficacy of existing communicative structures or work to develop new ones. Therefore, they have a unique understanding of the significance of certain developmental choices that may affect the efficacy and "user-friendliness" of a product or structure. Technical communicators should use this knowledge to participate in ongoing projects to create robust databases containing copyright holder information for works of various media that allow users who would not be able to benefit from copyright exceptions such as fair use to find copyright holders more easily.

Strategy 3: Advocate for Human Rights Interests in Global Solutions to Orphan Works

As noted earlier, the potential marginalization of developing and least-developed countries threatens their abilities to participate in the development of the international intellectual property framework. For practitioners, this possibly means making the sometimes difficult ethical choices of working for employers who support human rights interests (Herrington, 2010). For academicians, this could mean focusing on how technology and new means of communication affect societal constructs that may impact orphan works, particularly in how developed countries view the problems as compared to other, less-developed countries.

Strategy 4: Embrace Orphan Works

Technical communicators likely interact on a daily basis with works that are potentially orphaned. Therefore, technical communicators should advocate for a solution to orphan works that would help to address the fundamental disconnect in access that currently exists. Technical communicators should also initiate discussions about control of law, particularly in how it pertains to copyrighted works on the Internet for which there are no clear answers about control of law for search requirements for copyright holders and possible ensuing infringements. Practitioners who produce digital works in a work-for-hire setting could possibly attempt to initiate discussions with employers about the advantages of archiving older copyrighted materials so that society can still benefit from the ideas and expressions contained within them. These discussions will likely be difficult but could potentially improve the ability of society to access copyrighted works that would otherwise be lost.

CONCLUSION

Orphan works arise at an interesting intersection of technology and copyright law, and the local and the global. At both crossroads, technical communicators emerge as important players. Understanding the global significance of the orphan works problem requires consideration of how they affect the ability of technical communicators to work effectively. Furthermore, orphan works have an effect on democracy itself—how we interact to create new knowledge. Ultimately, it is these motivations, as well as the effects on speech and innovation, that should drive technical communicators to address the problems in their own work. By participating in the ongoing dialogue, technical communicators can potentially shape the development of a feasible orphan works solution, not only locally but globally. In doing so, technical communicators should keep in mind the potential effects on those who are limited in speech abilities, whether it is a portion of society or, in the case of developing or least-developed countries, entire nations. Through these discussions and the work of technical communicators, I strongly believe that there is hope for access, democracy, and the notions of self-actualization on the front of orphan works.

REFERENCES

Agreement on trade-related aspects of intellectual property rights, Apr. 15, 1994 (TRIPS). (1994). Marrakesh Agreement Establishing the World Trade Organization, Annex 1C, Legal Instruments—Results of the Uruguay Round, Vol. 31, 33 I.L.M. 1125, 1215.

Berne Convention for the Protection of Literary and Artistic Works (Paris Act 1971).

Birnhack, M. D. (2006). Global copyright, local speech. *Cardozo Arts and Entertainment Law Journal, 24,* 492–547.

Copyright Act of 1976, Pub. L. No. 94-553, 90 Stat. 2541, October 19, 1976.

Ginsburg, J. C. (2008). Recent developments in US copyright law: Part I—"Orphan" works. *Revue internationale du droit d'auteur, 217,* 1–21. Retrieved from http://lsr.nellco.org/columbia_pllt/08152/

Ginsburg, J. C., & Goldstein, P. (2005). *Reply comments on "orphan works" inquiry.* Retrieved from http://www.copyright.gov/orphan/comments/reply/OWR0107-Ginsburg-Goldstein.pdf

Herrington, T. K. (2001). *Controlling voices: Intellectual property, humanistic studies, and the Internet.* Carbondale, IL: Southern Illinois University Press.

Herrington, T. K. (2010). International fair use? *Technical Communication, 57,* 319–329.

Herrington, T. K. (2011). Copyright, free speech, and democracy: *Eldred v. Ashcroft* and its implications for technical communicators. *Technical Communication Quarterly, 20,* 47–72.

Huang, O. (2006). U.S. Copyright Office orphan works inquiry: Finding homes for the orphans. *Berkeley Technology Law Journal, 21,* 265–288.

Hughes, M., & Cohen Harrington, M. (2005, March 25). Comments of the Library of Congress in response to the Copyright Office Notice of Inquiry "orphan works" [Letter]. *Library of Congress.* Retrieved from http://www.copyright.gov/orphan/comments/OW0630-LOC.pdf

Nisbet, M. (2005). Orphan works notice of inquiry [Letter]. *Library Copyright Alliance.* Retrieved from http://www.copyright.gov/orphan/comments/OW0658-LCA.pdf

Okediji, R. (2000). Toward an international fair use doctrine. *Columbia Journal of Transnational Law, 39,* 76–175.

Patterson, L. R. (1987). Free speech, copyright, and fair use. *Vanderbilt Law Review, 40,* 2–66.

Proskine, E. A. (2006). Google's technicolor dreamcoat: A copyright analysis of the Google Book Search Library Project. *Berkeley Technology Law Journal, 21,* 213–239.

Société des auteurs des arts visuels et de l'image fixe v. SARL Google France, Tribunal de grande instance [T.G.I.] [ordinary court of original jurisdiction] Paris, May 20, 2008, 05/12117.

St.Amant, K., & Rife, M. C. (2010). Legal issues in global contexts: Reconsidering content in an age of globalization. *Technical Communication, 57,* 249–250.

Thompson, C. (2006). Orphan works, U.S. copyright law, and international treaties: Reconciling differences to create a brighter future for orphans everywhere. *Arizona Journal of International and Comparative Law, 23*(3), 787–852.

U.S. Register of Copyrights. (2006, January). *Report on orphan works.* Retrieved from http://www.copyright.gov/orphan/orphan-report-full.pdf

Van Houweling, M. S. (2005). Distributive values in copyright. *Texas Law Review, 83,* 1535–1579.

Wilson, A. N. (2009). Jet-setting orphan works: The transnational making available of works of unknown authorship, anonymous works, or lost authors. *Emory International Law Review, 23,* 784–820.

World Trade Organization (WTO). (2000, June 15). *United States—Section 110(5) of the US Copyright Act: Report of the panel.* Retrieved from http://www.wto.org/english/tratop_e/dispu_e/1234da.pdf

AFTERWORD

Global Challenges and the Value of Heuristics

Heidi A. McKee and James E. Porter

What strikes us most forcefully from reading the chapters in this collection is that legal issues in an online global context are complex, contradictory, confusing, and fluctuating. It is difficult to find legal guidelines that will apply with clarity across the types of contexts that technical communicators find themselves working in. As they write for global contexts—and aren't practically all communication contexts global ones at this time—technical communicators have to understand and navigate a messy, complex domain. These essays clearly drive this point home.

"What do we do about that?" is the important question that these essays ask. What principles, strategies, and procedures can technical communicators invoke to help them work with complex and multivarious contexts? The answer to that question, we think, has to focus, first, mainly on processes on the how and why—rather than by reference to specific "answers" found in legal mandates, and, second, on theoretical, ethical principles applied with contextual awareness rather than one-size-fits all application of set rules and concrete guidelines.

What do we mean by processes? We mean look to broad methods and inquiry heuristic rather than to prescriptions or algorithmic procedures. Nicole St. Germaine provides an example of this in her recommendation for technical communicators to use focus groups to test the effectiveness of healthcare information, particularly when translating healthcare information for speakers of different languages. That is a procedural and methodological recommendation about the how.

To some extent the choice to immerse oneself in the messy, rhetorical process of knowledge making is a difficult one for technical communicators, many of whom—take Brett Lunceford, for instance—desire clear answers and demand, as the solution to complexity, direct, clear, unambiguous, technically precise, and politically neutral language. Nice idea, but not realistic or practical. We're talking about a complex world here, a domain where semantic "clarity," whether in the use of terminology or in the understanding of legal questions, is an ideal seldom realized. Rather, technical communicators work in a domain where the linguistic and cultural diversity of audiences will make it impossible to achieve the kind of denotative clarity and technical precision that was so valued by John Locke in his theory of discourse (from *An Essay Concerning Human Understanding* [1690]), that became one of the defining principles of science writing espoused by the Royal Society, and that eventually became enshrined as a key defining stylistic revered in technical writing. We are advocates of clarity, too—yes, we think that people should "choose words carefully"—but that won't solve the global issues that technical communicators face.

All writing is, at some level, political, and all writing certainly is rhetorical, shaped by a myriad of factors including the complex intersections of legal systems and technological interfaces. Even in the realm of scientific measurement, we do not have uniformity or consistency, as Ronald Stone points out and as St. Germaine tells us: Chinese teaspoons and U.S. teaspoons are not the same, and this difference has important ramifications.

In a 2010 *Technical Communication* article on global issues (McKee & Porter, 2010), we argue that it is of course important for technical communicators to know "the law"—meaning the variety of laws and regulations that can apply when someone is doing global Internet research. But even more importantly, we argue, it is important for technical communicators to know how to ask questions, and to whom to pose those questions: "To seek answers in the ever-changing regulatory, legal, and cultural climate, technical communicators need to ask a range of questions, consult precedent cases, and seek advice from diverse audiences" (p. 282). In other words, our advice was procedural rather than substantive, and it focused more on case law, casuistic reasoning, and actual practices than on statutory law.

We would also point out that "the law" is not always the most important factor in communicating across global contexts. The law in a particular jurisdiction may not always correspond to actual practice or to cultural norms and expectations. What the laws says and what people actually do and think may be quite different, as we have pointed out (McKee & Porter, 2010) and as have contributors to this collection.

People's understandings about and expectations for privacy, shaped by the legal and cultural contexts in which they are immersed, vary widely across the globe, an issue Kraglung-Gauthier and Young explored in relation to online teaching, but that also applies to all online communications. Complicating any online communications is, as Marco Antonio Chávez-Aguayo detailed, the ever-changing nature of ICTs, how their evolving interfaces and everchanging affordances (not to mention their ever-changing privacy policies and terms of service) often outstrip legal and regulatory systems, leaving regulators, judges, lawyers, and technical communicators, not to mention plain old end users, in legal limbo. Chávez-Aguayo notes the need for more global cooperation and communication by governments and regulatory agencies—a point echoed by the many contributors in this collection who considered various aspects of the law.

A number of the chapters deal with the complexity and murkiness of international intellectual property laws and not just laws, but culturally different views of text, authorship, textual ownership, etc. (e.g., Pavel Zemliansky and Traci Zimmerman's chapter). As Annette Vee pointed out, in the United States we are in a realm where, increasingly, the default position toward ownership and control is moving toward corporate control and ownership of texts and authorship, and less control by the individual writer/author or by the consumer. (See also Lisa Potts' discussion of the loss of ownership for those who purchase Kindle texts through Amazon.) What is going to be the technical communicator's political position vis-à-vis these developments? Is the technical communicator's professional responsibility simply to "do what the boss wants" or what the corporation dictates, or what the client wants? Or as professionals do technical communicators have responsibilities to users, customers, consumers, maybe even to the public at large (see Porter, 2005)? From WIPO to WTO and at national and regional levels, IP law is continually evolving and changing, making it essential that technical communicators develop what Tatiana Batova described as "legal literacy."

Although focused on the complexity of technical translation, Batova's process-based recommendations for developing legal literacy are applicable across many contexts: technical communicators should continue "developing strategies for researching information on legal difference" across countries and cultures. Learning more about and developing the resources to share with others information about global legal issues in online contexts is essential. Technical communicators need to be informed about relevant policies and guidelines impacting particular communications. But, as Stone so aptly explained, "it may be useful to recognize that a standard or a guideline can serve as an appropriate starting point for guidance, but that it might not have been intended to be a final answer for a particular situation or scope of use." In other words, know your standards and guidelines, but be wary of relying on them too firmly as providing a "final, clear answer."

Because guidelines, laws, and standards are multiple (and often contradictory), technical communicators should seek more nuanced understandings, based on rhetorical principles, including the key principles of audience and context. In short: Know your audience and know your communicative context. Audiences are diverse, and contexts, especially global contexts, are complicated. More so than ever, technical communicators need to be wary not only of the assumption of linguistic homogeneity and cultural homogeneity, but also legal homogeneity. As the authors of this collection show, whether dealing with issues of translation, intellectual property, weights and measures, finance, or privacy, technical communicators are immersed in complex legal and global networks of laws, cultures, and language usage.

So what then should technical communicators do? Building from the many recommendations posed by contributors to this collection, we propose the development of heuristics for thinking through legal (and ethical) decisions. According to Richard Enos and Janice Lauer (1992), the ter, "heuristic comes from the Greek term *eurisko*, meaning 'to find out or discover.' The Latinized form of the term is *inventio*, which became a principal canon in the study of classical rhetoric. The term heuristic came to mean the *technai* or techniques within the *techne* of rhetoric that served to create effective discourse" (p. 204).

As we use the term, a heuristic is a set of open-ended questions, prompts, categories, memory devices, or visual grids to aid thinking, discovery, and deliberation (see McKee & Porter, 2009, 2010, 2012,; Porter, 2013).

The characteristic of an effective heuristic is that it is general and flexible enough to be applied across different contexts, and yet the prompts are specific enough to delineate the key issues and questions that communicators need to think about. For example, as we have proposed elsewhere (McKee & Porter, 2010), when considering intellectual property and the use of others' work, technical communicators need to ask certain kinds of questions about the use of that IP. What follows here is an excerpt from and adaptation of a larger heuristic—the first set of questions a researcher and communicator should ask about the material he/she is considering using:

Copyright Model for Technical Communicators: Questions to Consider

1. Identification of the material

- *Copying activity:* What materials will be copied for this research project?
- *Authorship:* Who is the author (or authors)?
- *Ownership:* Who is the copyright owner or owners?
- *Laws and regulations:* What legal jurisdiction applies to this material? (In what jurisdiction/s does the material arise?)
- *Explicit licensing:* What usage license, if any, is operative? In other words, what are the explicit directives and wishes of the author(s) and copyright owner(s)?
- *Implicit expectations:* What are the implied expectations for copying and redistribution of this material—for the author, the copyright owner, the community at large? Cultural attitudes, customs, mores. (McKee & Porter, 2010, p. 295)

Developed from the fundamentals of many countries' copyright laws and several international treaties these questions provide a set of exploratory prompts that are applicable across a wide-variety of contexts.

As contributors to this collection have shown, "the law" is a messy, moving target, constantly in a state of flux, open to various interpretations in different jurisdictions, and typically not up to date in regard to ICTs. For these reasons, technical communicators working in global contexts—and these days who isn't?—need to stay informed about legal and policy developments, government and institutional regulations, and the like. At the same time, they need to be wary of viewing any laws and regulations as providing definitive answers. In dialogue with colleagues, clients, target audiences, and legal and regulatory agencies, technical communicators need to employ informed critical judgment, based on an in-depth understanding of relevant legal, legislative, and regulatory frameworks and on the circumstances and details of the communicative contexts, as well as the cultural expectations and customs of the people and communities involved in the communication.

REFERENCES

Enos, R. L., & Lauer, J. (1992). The meaning of "heuristic" in Aristotle's *Rhetoric* and its implications for contemporary rhetorical theory. In S. P. Witte, N. Nakadate, & R. D. Cherry (Eds.), *A rhetoric of doing: Essays on written discourse in honor of James L. Kinneavy* (pp. 79-88). Carbondale, IL: Southern Illinois University Press.

Locke, J. (1690). An essay concerning human understanding, volume 1. Project Gutenberg. Retrieved from http://www.gutenberg.orglebooks/10615

McKee, H. A. (2011). Policy matters now and in the future: Net neutrality, corporate data mining, and government surveillance. *Computers and Composition.*

McKee, H. A., & Porter, J. E. (2009). *The ethics of Internet research: A rhetorical, case-based approach.* New York, NY: Peter Lang.

McKee, H. A., & Porter, J. E. (2010). Legal and regulatory issues for technical communicators conducting global Internet research. *Technical Communication, 57,* 282-299.

McKee, H., & Porter, J. E. (2012). The ethics of conducting writing research on the Internet: How heuristics help. In L. Nickoson & M. P. Sheridan (Eds.), *Writing studies research in practice: Methods and methodologies* (pp. 245–260). Carbondale, IL: Southern Illinois University Press.

Porter, J. E. (2005). The chilling of digital information: Technical communicators as publica dvocates. In M. Day & C. Lipson (Eds.), *Tech/Web: Technical communications and the World Wide Web in the new millennium* (pp. 243-259). Mahwah, NJ: Erlbaum.

Porter, J. E. (2011). Response: Being rhetorical when we teach intellectual property and fair use. In M. Courant Rife, S. Slattery, & D. N. DeVoss (Eds.), *Copy(write): Intellectual property in the writing classroom* (pp. 263–272). Fort Collins, CO: The WAC Clearinghouse and Parlor Press. Available at http://wac.colostate.edu/books/copywrite/

Porter, J. E. (2013). How does rhetoric theory inform the practice of technical communication? In J. Johnson-Eilola & S. A. Selber (Eds.), *Solving problems in technical communication.* Chicago, IL: University of Chicago Press.

Contributors

Editor Biographies

Kirk St.Amant is a professor of Technical and Professional Communication and of International Studies at East Carolina University.

Martine Courant Rife is a professor of writing and Coordinator of Accelerated Learning at Lansing Community College. Martine also maintains an active license to practice law in Michigan.

Author Biographies

Tatiana Batova is an Assistant Professor of technical communication at Arizona State University. Her research interests are in the areas of cross-cultural business, technical, and health care communication, content management, digital media, and rhetoric of technology. Tatiana has worked as a medical and pharmaceutical translator and writer, as well as a language specialist, interpreter, project manager, and consultant. In 2010 she received the Frank R. Smith Outstanding Journal Article Award from the Society for Technical Communication for "Writing for the Participants of International Clinical Trials: Law, Ethics, and Culture."

Marco Antonio Chávez-Aguayo (Ph.D., University of Barcelona) is an academic, with a multidisciplinary profile in Social Sciences and an artistic background in Music. He has been a lecturer and researcher at several universities in Mexico, Spain, and the UK. He has taught courses, done research and published work on various topics, ranging from Cultural Law, Cultural Policy and Cultural Management to Social Psychology, Sexuality and Education. He has a special interest in the application of new technologies. He is founder and director of the non-profit organization *Ópera Joven AC*, with which he is developing various cultural projects.

Martin S. Copenhaver is a Ph.D. student at the Operations Research Center at MIT. As an undergraduate student at Georgia Tech he became interested in issues in intellectual property policy and law after taking a class by Dr. TyAnna K. Herrington. Martin is interested in the influence of technology and rhetoric on access to information as well as theoretical questions regarding authorship, particularly as they play into other recent topics in copyright such as orphan works and harmonization.

TyAnna Herrington, JD, Ph.D., is a Professor at Georgia Tech. She specializes in intellectual property law and in international technical communication. Her books are in law and her articles treat issues in law and international communication, focused on digital learning. She serves on ATTW's Executive Committee as its Information Officer, is an Executive Advisory Board Member for CPTSC's *Programmatic Perspectives*, and is a member of the CCCC's Intellectual Property Task Force. Herrington served as a member of Georgia's State Board of Regents Copyright Committee and has delivered keynote, featured, and plenary addresses in venues including the NINCH Copyright Town Hall, CCCC, and CPTSC.

Wendy L. Kraglund-Gauthier is currently a Lecturer and Instructional Designer for the Faculty of Education at St. Francis Xavier University in Antigonish, Nova Scotia and a Ph.D. candidate at the University of South Australia. Her research and writing includes explorations of ways technology can be used in physical and virtual classrooms to promote collaborative, safe learning for students of differing ages and abilities. She is also interested in the impacts online technologies have on issues of privacy and on ethically conducting research online. Among the courses she has taught are Current Issues in Public Education, Sociology of Education, Inclusive Education, and 21st Century Teaching and Learning.

Brett Lunceford (Ph.D., Pennsylvania State University) is the editor of the *Journal of Contemporary Rhetoric* and the author of *Naked Politics: Nudity, Political Action, and the Rhetoric of the Body*. His research has appeared in such venues as *Communication Law Review, ETC: A Review of General Semantics, Explorations in Media Ecology, Media History Monographs, Northwestern Journal of Technology and Intellectual Property,* and *Review of Communication*, and in several edited volumes. He was previously a technical editor for Excite@Home.

Heidi A. McKee is an associate professor in the Department of English and an affiliate faculty member of the Armstrong Center for Interactive Media Studies at Miami University. She has co-edited three collections: *Digital Writing Research: Technologies, Methodologies, and Ethical Issues* (winner of the Computers and Composition Distinguished Book Award); *Technological Ecologies & Sustainability* (http://ccdigitalpress.org/ebooks-and-projects/tes); and *Digital Writing Assessment and Evaluation* (http://ccdigitalpress.org/ebooks-and-projects/dwae). With James Porter, she co-wrote *The Ethics of Internet Research: A Rhetorical, Case-based Process*.

James E. Porter is a Professor at Miami University, with a joint appointment in English and the Armstrong Institute for Interactive Media Studies. He also serves as Director of the American Culture & English program.

Liza Potts is an assistant professor in the department of Writing, Rhetoric, and American Cultures at Michigan State University. She is a senior researcher at WIDE Research. Her research interests include technologically-mediated communication, experience architecture, and participatory culture. Her work has appeared in *Technical Communication Quarterly*, *Journal of Business and Technical Communication*, and *Technical Communication*. Her new book, *Social Media in Disaster Response*, focuses on how everyday people communicate in digital spaces during times of crisis.

Nicole St. Germaine is an assistant professor of technical and business writing at Angelo State University, as well as a freelance health and legal writer. Her research interests include technical communication for a Mexican-American audience and technical communication in the health fields. She is the two-time recipient of the Distinguished Article award from the journal *Technical Communication.*

Ronald L. Stone is a communications consultant in Northern California who has also been actively involved in developing AAT ICAS standards and programs for the nonprofit Alliance for the Advancement of Technology (AAT). Ronald produced the metrication month poster for 2009, usability comparison of calendar date formats, for AAT. At the STC 2011 Annual Conference, Ronald discussed a study of strategies for organizing global and local formats. Ronald has also contributed newsletter articles to STC and IEEE communities or groups.

Annette Vee received her Ph.D. from the University of Wisconsin-Madison and is an Assistant Professor of English at the University of Pittsburgh, where she teaches undergraduate and graduate courses in literacy, technology, writing, and digital composition. Her research explores digital intellectual property debates as well as intersections between alphabetic and computer code writing in historical contexts. She has published work in Computers and Composition, Enculturation, Literacy in Composition Studies and JAC. Her current book project argues that the historical and social perspective of New Literacy Studies can help us understand the technical and social affordances of computer code writing.

David C. Young is an Associate Professor in the Faculty of Education at St. Francis Xavier University in Antigonish, Nova Scotia. His research is focused on the broad topic of educational administration and policy. More particularly, his current writing deals with issues surrounding law and education. Among the courses he has taught are Psychological Foundations of Education, Educating Exceptional Students, Social Foundations of Education, Educational Policy, Educational Research, School Administration and Education Law.

Pavel Zemliansky is an associate professor in the Department of Writing and Rhetoric at the University of Central Florida, where he also directs the UCF's Writing Across the Curriculum Program. His research interests include professional communication, especially in international contexts, WAC/WID, and writing pedagogy. His publications include books and book chapters on professional communication, writing studies, and rhetoric, as well as articles in such journals as *IEEE Transactions on Professional Communication*, *Kairos*, and others.

Traci A. Zimmerman is currently an associate professor in the School of Writing, Rhetoric and Technical Communication at James Madison University, in Harrisonburg, Virginia. She teaches classes in authorship and intellectual property; language, law and ethics; rhetorical theory and literacy; and linguistics. Her current research is informed by her involvement with the Conference on College Composition and Communication (CCCC) Intellectual Property Caucus and IP committee and is inspired by her interest in advocating for greater development and innovation in open access publishing and open education.

Index